CRYSTAL PHYSICS

Macroscopic Physics of Anisotropic Solids

*Modern Physics Monograph Series**
Editor: FELIX M. H. VILLARS

Published:

(1) Robert T. Schumacher
   Introduction to Magnetic Resonance: Principles and Applications, 1970

(2) Robert J. Finkelstein
   Nonrelativistic Mechanics, 1973

3 Hellmut J. Juretschke
   Crystal Physics: Macroscopic Physics of Anisotropic Solids, 1974

*Note for librarians/bibliographers:
Starting with the volume issued in 1974, books in this Series will be numbered.

CRYSTAL PHYSICS

Macroscopic Physics of Anisotropic Solids

Hellmut J. Juretschke
Polytechnic Institute of New York

1974
W. A. BENJAMIN, INC.
ADVANCED BOOK PROGRAM
Reading, Massachusetts

London · Amsterdam · Don Mills, Ontario · Sydney · Tokyo

CODEN: MPMSC

Library of Congress Cataloging in Publication Data:

Juretschke, Hellmut J.
  Crystal Physics.

  (Modern physics monograph series)
  1. Solids.   2. Anisotropy.   3. Crystallography, Mathematical.   I. Title

  QC176.J83   548'.8   73-17083

  ISBN 0-8053-5102-7 (hardbound)
  ISBN 0-8053-5103-5 (paperback)

Copyright ©1974 by W. A. Benjamin, Inc.
Published simultaneously in Canada.

All rights reserved. No part of this publication may be reproduced, stored in a retrieval system or transmitted in any form or by any means, electronic, mechanical, photocopying, recording, or otherwise, without the prior written permission of the publisher,
W. A. Benjamin, Inc., Advanced Book Program, Reading, Massachusetts 01867, U. S. A.

Manufactured in the United States of America

ABCDEFGHIJ-MA-7987654

Contents

Editor's Foreword	xiii
Preface	xv

CHAPTER 1

Anisotropy   1

CHAPTER 2

Crystal Symmetry   4

2.1.	Macroscopic Properties Exhibit Point Group Symmetry	5
2.2.	Lattices in Two Dimensions	5
2.3.	Group Properties of Symmetry Operations	6
2.4.	Point Groups in Two Dimensions	9
2.5.	Lattices and Point Groups in Three Dimensions	10
	Problems	14
	Bibliography	15

CHAPTER 3

Mathematical Description of Crystal Properties   16

3.1.	The Symmetry of Physical Properties	16
3.2.	Restrictions on Matter Tensors	17
3.3.	Orthogonal Transformations	18
3.4.	Consecutive Transformations and Matrix Representations of the Symmetry Group	20
3.5.	Transformation Properties of Tensors	21
3.6.	A Practical Rule for Carrying out Tensor Transformations	22
3.7.	Determination of the Independent Constants	24
	Problems	26
	Bibliography	28

CHAPTER 4

Tensor Symmetry and Linear Vector Spaces — 29

- 4.1. Tensor Invariants — 29
- 4.2. Tensor Symmetry in Isotropic Materials — 31
- 4.3. Invariant Subspaces of Tensor Components — 34
- 4.4. Invariants of Transformations — 38
- 4.5. Crystal Symmetry — 42
- Problems — 45
- Bibliography — 47

CHAPTER 5

Electric Polarization — 48

- 5.1. Formulation of the Interaction — 48
- 5.2. Symmetry Considerations — 50
- 5.3. Depolarization Fields and Shape Anisotropy — 52
- 5.4. Measurement of Dielectric Constants — 54
- 5.5. Potential Distributions — 57
- 5.6. Comments on Magnetic Systems — 58
- Problems — 58
- Bibliography — 60

CHAPTER 6

Magnetic Symmetry — 61

- 6.1. Time Reversal — 62
- 6.2. An Example of Magnetic Symmetry — 63
- 6.3. The Magnetic Groups — 65
- 6.4. Mathematical Description of Magnetic Properties — 66
- 6.5. Some Applications — 68
- Problems — 69
- Bibliography — 70

CHAPTER 7

Electrical Conduction   **71**

7.1.	The Symmetry of Irreversible Processes	71
7.2.	Anisotropic Conduction	74
7.3.	Potential Distributions in Anisotropic Conductors	76
7.4.	Four Probe Measurement of Conductivity	79
7.5.	Electrical Transport in a Magnetic Field	81
7.6	Symmetry of the Hall Effect	82
7.7.	Potential Distributions Including the Hall Effect	84
7.8.	Transport in Magnetic Materials	86
	Problems	88
	Bibliography	90

CHAPTER 8

Thermoelectricity   **91**

8.1.	Currents and Driving Forces	92
8.2.	Thermoelectric Heat Generation	94
8.3.	Thermoelectric Effects	95
8.4.	Thermoelectric Relations	99
	Problems	100
	Bibliography	103

CHAPTER 9

Crystal Optics   **104**

9.1.	General Properties of Normal Modes	105
9.2.	Fresnel Formulas	106
9.3.	Uniaxial Crystals	108
9.4.	Biaxial Crystals	111
9.5.	Coupling to an Isotropic Medium	113
	Problems	115
	Bibliography	119

CHAPTER 10

Second-Order Optical Effects   **120**

10.1.	Optical Activity: Symmetry Considerations	120
10.2.	Optical Activity: Approximate Dispersion Relations and Normal Modes	123
10.3.	Effects of Static Magnetic or Electric Fields	126
10.4.	Free Energy Formulation of Optical Effects	128
	Problems	130
	Bibliography	133

CHAPTER 11

Elasticity   134

11.1.	The Elastic Parameters	134
11.2.	Transformation Properties and Symmetry Considerations	137
11.3.	Isotropic and Polycrystalline Media	140
11.4.	Determination of Elastic Constants by Static Methods	142
11.5.	Isothermal and Adiabatic Constants	144
11.6.	Elastic Waves	145
	Problems	148
	Bibliography	153

CHAPTER 12

Piezoelectricity   154

12.1.	Thermodynamics of Piezoelectricity	154
12.2.	Transformation Properties of Piezoelectric Tensors	157
12.3.	Elastic Waves in Piezoelectric Crystals	158
12.4	Piezoelectric Effects in Electromagnetic Waves	160
12.5.	Optical Phonons and Polaritons	161
	Problems	165
	Bibliography	168

CHAPTER 13

Formulation of Higher Order Interactions   169

13.1.	Various Fourth Rank Tensors	169
13.2.	Fifth and Sixth Rank Tensors	171
13.3.	Isotropic and Cylindrical Symmetry	174
13.4.	Polycrystalline Media	175
13.5.	Third Order Elastic Coefficients in Cubic, Isotropic, and Polycrystalline Media	177
13.6.	Polycrystalline Axially Symmetric Media	182
	Problems	185
	Bibliography	190

Contents

APPENDIX 1

The 32 Crystallographic Point Groups and Their Symmetry Operations 191

Symmetry Elements 191
Symmetry Groups 191
Symmetry Operations 192

APPENDIX 2

Generating Elements of the 32 Point Groups 194

APPENDIX 3

Linear Combinations of Tensor Components Transforming in Invariant 196
Subspaces Under the Covering Operations of the Rotation Group in Three
Dimensions

APPENDIX 4

Characters of Representations of the Group of Rotations in Three 200
Dimensions

APPENDIX 5

Irreducible Characters and Character Tables of the 32 Crystallographic 205
Point Groups

APPENDIX 6

The Magnetic Point Groups and Their Symmetry Elements 210

SELECTED LITERATURE 214

INDEX 217

Editor's Foreword

Education in physics is going through a phase of rapid evolution. On the frontier of the field new information is literally pouring in, new perspectives are opening up, and new concepts are emerging. For the student, the distance to be covered from freshman year to graduate research work appears to be ever expanding.

Professional education in physics therefore must deal with the very real problem of the need for thoughtful condensation of the material presented, and with the question of what may and should reasonably be achieved in the years between the introductory and the research level.

It is generally agreed, on the one hand, that a thorough presentation of the fundamentals of both classical and quantum physics is essential. On the other hand, there is the understandable desire to let the student participate in the excitement offered by the many interesting new developments in all fields of physics. The discussion of such new topics gives the student an opportunity to see the actual growth process of science: new experiments, new techniques, and the attempts to relate new results to existing or emerging theoretical views. The study of the well-established traditional subjects of physics appears to lack these exciting aspects, and to offer little room for the display of creativity, except as historical fact.

It has at last been recognized that this need not entirely be so; that, in fact, the close ties between the traditional and the modern can be exploited to establish contacts between the classical subjects and current endeavors: classical mechanics and space navigation, wave optics and radar interferometry, or holography, astrophysical applications of classical electromagnetism and hydrodynamics, statistical mechanics as applied to biopolymers, or phase transitions, and so forth. To develop such links wherever they exist, and to put the essential parts of the traditional subjects into a modern perspective is an urgent and rewarding task.

On the undergraduate level, the recent burgeoning of such introductory texts as the Feynman lectures, the Berkeley physics course, and the Massachusetts Institute of Technology introductory physics texts bear witness to the interest that has been aroused by the problem of bringing the fundamentals of physics to the undergraduate in a novel way. This new series of MODERN PHYSICS MONOGRAPHS intends to continue this process at the more advanced level. It will present material for the upper-level undergraduate and introductory graduate courses. At this level there will be, on the one hand, courses of a specialized nature, with the purpose of giving the student an introduction to the great diversity of topics of physical science, from particle physics to nuclear, atomic, solid state, plasma and astrophysics; while, on the other hand, the traditional topics of the undergraduate sequences will be deepened and extended, and their interrelations more clearly established. We hope that the MODERN PHYSICS MONOGRAPH Series will help to give the lecturer in the field additional flexibility in choosing his course material and, if he is inclined to experiment, allow him to introduce into his course topics not generally covered in standard textbooks. In addition, the student will have access to a variety of collateral reading material.

For these very reasons, the books of this series are *not* intended to be textbooks, but rather monographs; that is, works that cover a more restricted area in a space of approximately 100 to 250 pages. They contain problems with and without answers, and could either supplement existing texts or be used in groups as a replacement for a single text.

The present volume deals with the macroscopic physics of anisotropic solids. It is a good example of a subject largely bypassed in introductory courses, and often neglected even in advanced presentations of solid state physics, which nowadays very much emphasizes the microscopic description of phenomena. Yet, all macroscopic manifestations of interactions on the microscopic level fit into a general framework, determined by crystal symmetry and the nature of the applied interaction.

In this monograph, the author admirably succeeds in showing that this general framework can be established and presented by quite elementary means, making no demands on the student beyond an understanding of elementary group theory and linear algebra. It is thus well within the reach of an advanced undergraduate.

We find it also worth mentioning that a systematic study of macroscopic relations reveals how much experimental work remains to be done in this field. As the author points out, in Chapter 1: "The systematic exploration of highly anisotropic crystals is just beginning, and some of the effects predicted in such crystals remain to be discovered experimentally."

We may hope then that this monograph will not only offer some rewarding reading, but perhaps stimulate some experimental investigations as well.

FELIX M. H. VILLARS

Preface

This book on the macroscopic physics of anisotropic media developed out of a series of introductory lectures for a course in crystal dynamics. If the approach to crystal dynamics follows a text such as Born and Huang's *Dynamical Theory of Crystal Lattices*, the macroscopic properties of solids deriving from interactions at the atomic level automatically exhibit the full anisotropy of single crystals. In order to sort out the results obtained in this fashion and to assign them to the traditional effects, such as magnetoresistance, piezoelectricity, or optical activity, it is helpful to know beforehand the possible form of these effects in crystals of different symmetry and under various experimental conditions. This knowledge follows from a purely macroscopic formulation of the physics of anisotropic media. Such a formulation, built on the combined symmetries of physical processes and of crystal structure, effectively establishes the framework within which all microscopic theories must operate. Of course, crystal physics at the macroscopic level is equally important in its own right in dealing with the great variety of new phenomena that become possible once the restrictive assumption of the isotropy of matter, under which most physics is traditionally taught, is removed.

In this book we sample a number of areas of classical physics where the complication of matter anisotropy plays an important first-order role. Each topic includes a review of the basic physics of the interaction in question, a discussion of the symmetry restrictions governing the effect, and some applications that emphasize the consequences of anisotropy or that illustrate the generalization of well-known concepts to anisotropic media. The choice of topics is conventional, but the selection of points taken up for each topic and the emphasis on certain applications were governed largely by the questions my students and I wanted to have clarified. While the book is self-contained, it relies on other texts in crystal physics, such as Nye's *The Physical Properties of Crystals*, for more detailed discussion of standard material and for the well-known tables of tensor coefficients of the more familiar effects. Consequently, there is room in practically every chapter for more extensive examination of some consequences of matter anisotropy and for additional applications.

The mathematical apparatus, consisting primarily of simple tensor algebra and elementary group theory, is introduced in the early chapters developing a quantitative formulation of symmetry. Throughout the text, the emphasis is on using these mathematical tools in the simplest form adequate for our needs, and on introducing additional aspects only as they are needed in connection with some specific application. Yet we also touch on topics, such as the role of linear invariants in the discussion of the properties of polycrystalline media, that call attention to additional uses of group theory as a practical tool.

The problems at the end of each chapter are an integral part of the text. Some fill in details of derivation skipped in the general discussion, others illustrate specific applications, or extend the arguments of the text to more complex or wider-ranging situations. Wherever the answer to a problem may be of more immediate interest to a worker in crystal physics than the details of its derivation, the problem has been formulated so as to display the result. On the other hand, problems of the same type that could be asked about each separate topic are not included systematically. They can easily be posed by the reader, as can many additional examples that emerge naturally upon studying the text.

References to other treatments of the subject matter are given at the end of each chapter. In addition there is a list of selected readings intended to give a small sampling of the original literature. Finally, the Appendices contain the material necessary to extend the approach of this text to other interactions in crystals, already known or yet to be discovered.

It is impossible to acknowledge the many points of detail which I have freely adapted from textbooks and articles in the field. In addition, this book owes much to my students and to my teachers. I am grateful for all the help and criticism received in its preparation. Above all, I want to thank P. P. Ewald for his most important role in this project. As my former department head, he gave me the assignment to develop and present a course in crystal dynamics. His own interests in this field have been an incomparable source of encouragement, and through his wisdom and enthusiam I came to share in the appreciation of the accomplishments of the great crystal physicists.

Mrs. Mollie Bartels, Mrs. Kathy Schubler, and Mrs. Migdalia Alvarez typed the manuscript with diligence, patience, and an enormously helpful attention to detail. Finally, my admiration and gratitude go to my family for so patiently giving of space, time, and warmth when the book had to be written.

HELLMUT J. JURETSCHKE

CHAPTER 1

Anisotropy

In the traditional theories of physics the properties of matter enter through a set of numerical factors, with each factor characterizing the response of the material to an applied force or field. These numbers, such as the dielectric constant ϵ, the index of refraction n, or the thermal conductivity K, are taken as *scalars*, since the response of the material is *linear* and *isotropic*. Isotropy means that all directions in the solid are equivalent. It implies, for example, that the velocity of light is independent of direction of propagation, or that a thermal current must flow against the gradient of temperature. The equivalence of all directions means that the solid has full rotational symmetry.

The representation of the properties of matter by scalar coefficients is insufficient as soon as we deal with single crystals. The response of a crystal to external forces or fields depends not only on the magnitude of these influences but also on their orientation relative to reference directions proper to the crystal. Crystals are *asymmetric*. On the other hand, being built on a periodic structure at the atomic level, crystals must exhibit some of the symmetry of such a structure. In fact, most crystals are *anisotropic*; that is, they show limited symmetry. In such restricted symmetry, the set of all directions breaks up into subsets that contain all the directions equivalent under a finite and closed group of transformations of the reference directions of the crystal. As a consequence, the scalar characterizing a physical property of an isotropic solid is replaced in a crystal by a finite set of coefficients that obey the transformation properties of a *tensor*. The simple isotropic formulation is recovered only in *polycrystalline* materials composed of randomly oriented single crystallites of sufficiently small dimensions so that on the scale of macroscopic interactions the material appears homogeneous.

While the scalar constants characterizing a polycrystalline medium allow a simple isotropic theory of the interaction of the solid with external influences, they contain only rather incomplete information concerning the properties of the single crystallites making up the medium. To obtain full information, we must study single crystals, and to interpret the response of crystals to various applied forces, we must formulate the theories of classical physics for anisotropic media.

We expect that as a result of its anisotropy the response of the medium will be more varied, and that it will also include new effects that become possible specifically because the medium is anisotropic. As a matter of fact, the richness in response becomes so large that many of the macroscopic aspects of anisotropy have not yet been fully tested or exploited. The number of anisotropic crystals for which full and reliable knowledge of the simplest material constant tensors is available is surprisingly small.

The generalization of macroscopic physics to include anisotropic interactions is, in principle, straightforward. In practice, it can become tediously bogged down with extensive algebra and complex expressions whose physical import is not easily discerned. To minimize these difficulties it is essential to formulate the theory in compact and consistent notation that includes from the outset all the symmetries of the interaction under discussion. This is achieved by employing tensors that are simplified by systematically applying the methods of elementary group theory to take advantage of these symmetries.

The symmetries entering into the theory are of two kinds. There are inherent symmetries arising out of the nature of the forces or fields involved in the interaction. They manifest themselves in conservation laws, reciprocity relations, and the like. In addition, there are symmetries involving space and time connected with the specific anisotropy of the medium in which the interaction takes place. Both kinds of symmetry can have a powerful influence on defining the allowed structure of matter tensors.

In fact, the formulation of anisotropic physics based on these principles of symmetry determines primarily the necessary framework of all possible interactions. It separates allowed and forbidden effects, and it specifies the form of the allowed effects in a crystal of given symmetry without making any prediction as to whether a particular crystal of this symmetry shows a strong or weak response, or any response at all. Interesting physics, of course, concentrates on those crystals showing large effects. Our approach focuses on the preliminary aspects of determining the classes of crystals to which the search for any particular effect can be restricted, and the form of the response by which this effect is to be identified.

This aspect of crystal physics covers a sufficiently large ground to justify development in its own right. Surprisingly enough, it is not a closed subject. The systematic exploration of highly anisotropic crystals is just beginning, and some of the effects predicted in such crystals remain to be discovered experimentally. On the other hand, the recent history of crystal physics includes instances of "forbidden" effects that were found to be "allowed" after more careful examination of the inherent symmetries governing the interaction or of the structure symmetries of crystals. Crystal physics has had its share of "symmetry violations" and of the new worlds opened up by each such discovery. There is good reason to believe that the full intricacy of the interactions possible in crystals is yet to be discovered.

CHAPTER 2

Crystal Symmetry

A crystal is a solid whose local properties on a microscopic or atomic scale are periodic in three dimensions. This *translational symmetry* follows from the periodic spatial arrangement of the atoms or molecules making up the solid. The basic configurational unit of the crystal, the *cell*, generates the total solid by repeating upon displacement along integral multiples of three independent *basis vectors*. The content of the cell (the *basis*) and the network of points defining the location of the repeated cell centers (the *lattice*) together make up a specific *crystal structure*.

In addition to translation, a crystal may possess symmetry under *rotation*, *reflection*, and more complex covering operations combining these with translations. This additional symmetry results from the symmetry of the arrangement of atoms within the cell (determined by their manner of interaction, such as bond lengths and bond angles), taken together with the symmetry of the lattice, and the specific positioning of the cell on the lattice. Compatibility requirements between the symmetries of cell, lattice, and cell positioning lead to the result that the number of distinct types of spatial arrangements of crystals is finite, and actually relatively small. All such possible arrangements are described by the *crystallographic space groups*. The specification of the symmetry of any one arrangement requires listing all the basic covering operations (*symmetry elements*) under which the structure goes over into itself, together with the locations of points, axes, or planes with respect to which the covering operations are carried out.

The space groups list all symmetries realizable at the microscopic level where the location of each and every atom is specified. For our purposes this description is, in fact, too complete; therefore, we are interested in a systematic reduction of the totality of microscopic symmetries to those which suffice to describe the macroscopic behavior of anisotropic solids.

2.1 MACROSCOPIC PROPERTIES EXHIBIT POINT GROUP SYMMETRY

The symmetry of the local environment in a crystal varies from point to point, and the local reaction to an external influence is more pronounced at some points in each crystal cell than at others. However, external influences that are uniform on a scale covering many characteristic repeat distances of the crystalline periodic arrangement cause the crystal cell to react essentially as a whole. The formal description of this macroscopic interaction must be independent both of the location of an origin of coordinates and of the detailed location of the various symmetry elements within a crystal cell. Hence, for macroscopic effects we consider all such symmetry elements to act at the same point and, indeed, at all points of the infinite solid. The crystalline solid therefore can be approximated by a *continuous solid with uniform anisotropic point properties*.

The various possible sets of symmetries associated with such a solid are the *crystallographic point groups*. They are derivable starting from the crystal space groups by condensing all symmetries that differ only by the relative location of their symmetry elements within the unit cell. Alternatively, the crystallographic point groups can be obtained by starting from the symmetry of crystal lattices and then including symmetric structures attached to each lattice point. This latter approach is much more direct. It is developed in the following sections for groups in two dimensions and then applied to groups in three dimensions.

2.2. LATTICES IN TWO DIMENSIONS

Two independent basic translations $\mathbf{a}_1$ and $\mathbf{a}_2$ define all lattice points P of a two-dimensional net (n_1, n_2)

$$\mathbf{P} = \Sigma n_i \mathbf{a}_i \tag{2-1}$$

The symmetry of such a net of points is specified by enumerating the totality of distinct symmetry elements or spatial symmetry operations that, when applied to the lattice about a lattice point, take the lattice into itself. The complete set of symmetry elements depends on the relative direction and magnitude of the two basic vectors.

All plane lattices possess at least two symmetry elements: (a) the identity operation 1, which consists of leaving all points of the lattice unchanged; (b) the inversion operation 2, which inverts all lattice points through the origin $\mathbf{P} = 0$. (This operation is given the symbol 2 because inversion in two dimensions is equivalent to a twofold rotation about an axis normal to the plane.)

The additional possible symmetry elements are *rotations* and *reflections*. Rotation elements are labeled by the number of repeats of the rotation angle that make a full circle. Thus, 2 stands for the rotation through 180°. Reflection elements are labeled by m to indicate that they correspond to mirror operations.

Fundamental rotations of plane lattices about a normal axis must be *even* in order to be compatible with the common symmetry 2. On the other hand, more than sixfold symmetry generates points on the circumference of a circle closer together than the initial basic spacing specified by the radius of the circle. Hence, beyond 2 the only additional rotational symmetry elements are 4 and 6. Of course, repeated application of these generates other, and usually distinct, symmetry elements. Thus, the six symmetry elements associated with the generating element 6 are

$$6, 6^2 = 3, 6^3 = 2, 6^4 = 3^2, 6^5, 6^6 = 1$$

This list indicates that the element 3 appears as part of the family of elements generated by 6 even though it cannot exist alone as a lattice symmetry. Furthermore, the list includes as distinct elements rotations through the same angle but in opposite sense, as for example 6 and 6^5.

The *order* of an element is the exponent to which the symmetry element must be raised to generate the identity (6 is an element of order 6: $6^6 = 1$).

Reflections m are represented by mirror planes containing the normal axis. They can either interchange axes or turn a single axis into its negative. They are elements of order 2: $m^2 = 1$.

Rotations other than 2 or reflections occur in lattices defined by special relations between a_1 and a_2. Four such special relations exist. Hence there are five distinct types of plane lattices. These five lattices are listed in Table 2-1 according to the relations between the magnitudes and directions of their basis vectors. Table 2-1 also lists the totality of the distinct symmetry elements of each lattice. The accompanying figures clarify the lattice geometry and the disposition of the mirror planes. In lattices B, C, and D the primed mirror plane m' is at right angles to the unprimed one. In lattice E, m' and m'' are at 60° and 120° with respect to m.

2.3. GROUP PROPERTIES OF SYMMETRY OPERATIONS

Each symmetry element defines a symmetry operation taking the lattice into itself. If two symmetry operations associated with a lattice are applied consecutively, we speak of the *product* YX of two operations X and Y (operation X is followed by operation Y). Since the product operation also takes the lattice from its initial standard configuration into a final indistinguishable configuration, it must be equivalent to a single operation. The equivalent operation is easily

Table 2-1
Lattices in Two Dimensions

Lattice type	Geometry	Symmetry elements
A. $a_1 \neq a_2$, ϕ arbitrary		1, 2
B. $a_1 \neq a_2$, $\phi = 90°$		$1, 2, m_1, m_1'$
C. $a_1 = a_2 = a$, ϕ arbitrary		$1, 2, m_2, m_2'$
D. $a_1 = a_2 = a$, $\phi = 90°$		$1, 2, 4, 4^3$ m_1, m_1', m_2, m_2'
E. $a_1 = a_2 = a$, $\phi = 60°$		$1, 2, 6, 3, 3^2, 6^5$ $m_1, m_1', m_1'',$ m_2, m_2', m_2''

found by applying the two consecutive symmetry operations to the standard set of basis vectors, and then matching the end result with the known results of a single symmetry operation. For example, Table 2-2 shows the effect of applying

Table 2-2
Application of Symmetry Operations of Lattice B to the Standard Set of Basis Vectors

Operation	Initial configuration	Final configuration
1		
2		
m_1		
m_1'		
2		
m_1'		

the symmetry operations of lattice B to the standard set of basis vectors of this lattice. A second symmetry operation is then applied to the configuration resulting from the first operation. By rotating the configuration of m_1 through 180° it follows immediately from Table 2-2 that $2m_1 = m_1'$. Similarly, reflecting the configuration of 2 in a vertical plane m_1' leads to the result that $m_1^{2'} = m_1$. The same process can, of course, be repeated for products of three or more symmetry operations.

The reduction of products to single operations can be completely summarized in a *multiplication table*. In giving the product of any two symmetry operations, either order is included because the two operations may not commute. For the lattice B the multiplication table takes the form of (2-2). Such a table fully defines

$$
\begin{array}{c|cccc}
 & \multicolumn{4}{c}{X} \\
 & 1 & 2 & m_1 & m_1' \\
\hline
1 & 1 & 2 & m_1 & m_1' \\
2 & 2 & 1 & m_1' & m_1 \\
m_1 & m_1 & m_1' & 1 & 2 \\
m_1' & m_1' & m_1 & 2 & 1 \\
\end{array}
\qquad (2\text{-}2)
$$

with Y labeling the rows.

the multiplication operations for symmetry elements. It implies that all symmetry operations and their products form a small, self-contained, closed set. This set includes the identity 1 and permits defining the inverse X^{-1} of any operation X by the rule $X^{-1}X = XX^{-1} = 1$.

The set of symmetry elements of any lattice together with the properties just described defines its *symmetry group*. Groups are usually labeled by enclosing their *generating elements* in parentheses. The generating elements are those elements that by repeated application among themselves generate the whole group. In terms of this convention, the five lattices of Table 2-1 have the symmetry groups: A, (2); B, $(2m_1)$; C, $(2m_2)$; D, $(4m_1)$; E, $(6m_1)$.

Symmetry groups not only characterize lattices, but label all two-dimensional point symmetries. The task of determining these is, in fact, that of enumerating all distinct two-dimensional point symmetry groups. In this connection one property of groups is particularly relevant. *Groups may contain among their elements subsets that form groups of lower order*. For example, the group $(2m)$ corresponding to lattice B or C contains the following distinct subgroups: $(1) = 1$; $(2) = 1, 2$; $(m) = 1, m$; $(2m) = 1, 2, m, m'$. Each of these sets of elements forms a group with its own multiplication table. It is customary to include in this list of subgroups both the trivial group (1) and the full group $(2m)$; however, we do not consider the subgroup $1, m'$ distinct from $1, m$. Except for $(2m)$, these subgroups do not describe the full symmetry of lattice B. Rather, they define all closed subsets of symmetry operations compatible with this lattice.

2.4. POINT GROUPS IN TWO DIMENSIONS

The results of Section 2-3 are sufficient to construct all point groups in two dimensions. These groups are, of course, well known, and have been derived before by reasoning ranging from explicit geometric construction to rather abstract algebra. Our derivation relies on some general ideas of elementary group theory.

So far, we have only discussed the symmetry of lattices. To obtain a crystal, we populate all lattice points with a specific basis that has its own symmetry. The symmetry of the resultant crystal depends on the symmetry of the lattice, the symmetry of the basis, and the amount of agreement between the reference points of the symmetry elements of both.

The pivotal rule for determining the crystal symmetry in these circumstances is based on the following argument. *Regardless of the choice of basis, the symmetry of the crystal must belong to one of the subgroups of the lattice symmetry group.* Only such groups are at all compatible with the lattice arrangement. Furthermore, these subgroup symmetries are realized only if they also are contained in the symmetry of the basis. Otherwise, the crystal symmetry is reduced to the trivial group (1).

Based on this reasoning the number of distinct group symmetries in two dimensions is obtained simply by enumerating all the possible distinct subgroups of the plane lattice symmetries. The list of groups and their elements in Table 2-3 is the result. In this list the groups are associated with the least symmetric lattice in which they first appear. Once the groups have been established, they can, of course, describe a crystal based on any of the lattices from which they are derivable, so that there is no longer a direct association of a symmetry group with any one lattice. Furthermore, the final list of point groups also does not distinguish between symmetry elements m differing only by orientation of their plane, such as m_1 and m_2 (as long as they do not appear together). Consequently, for instance, lattices B and C are macroscopically indistinguishable.

Table 2-3
Point Groups in Two Dimensions

	Group	Elements
A	(1)	1
	(2)	1, 2
B	(m)	1, m
	(2m)	1, 2, m, m'
D	(4)	1, 4, 2, 4^3
	(4m)	1, 4, 2, 4^3, m_1, m_2, m_1', m_2'
E	(3)	1, 3, 3^2
	(3m)	1, 3, 3^2, m, m', m''
	(6)	1, 6, 3, 2, 3^2, 6^5
	(6m)	1, 6, 3, 2, 3^2, 6^5, m_1, m_1', m_1'', $m_2 m_2', m_2''$

The ten point symmetry groups listed in Table 2-3 fully describe the symmetry of all macroscopic properties of strictly two-dimensional crystals (as distinct from single layers of real crystals). Interest in these two-dimensional groups is not entirely academic, however, since they represent the symmetry of those three-dimensional crystals for which the third dimension does not introduce additional symmetry elements.

2.5. LATTICES AND POINT GROUPS IN THREE DIMENSIONS

The point groups of real interest are those describing the symmetry of three-dimensional crystals. Their derivation is straightforward in terms of the approach of the two preceding sections. First we construct the distinct three-dimensional lattices and define their symmetries. Then all point groups are formed by the totality of groups and subgroups of the lattice symmetries.

Three-dimensional lattices can be built from the lattice planes of Table 2-1. The symmetry introduced by the third basic translation $\mathbf{a}_3$ is specified by the direction and the order of a rotation axis at an angle to the plane in question. Just as in the two-dimensional case, special relations occur for special angles and magnitudes of $\mathbf{a}_3$. The universal symmetry of all three-dimensional lattices is the inversion through a lattice point $\bar{1}$. The list of lattices and their construction follow. (We use standard notation for describing symmetry elements as they arise. The complete specification of these symmetry elements is given in Appendix 1).

1. *Triclinic System.* If $\mathbf{a}_3$ is of arbitrary length and not normal to any of the planes A, B, C, D, E, there is no rotation axis outside the planes. The three-dimensional network of points exhibits only inversion symmetry. All such networks form a single lattice type, with group $(\bar{1}) = 1, \bar{1}$. A sample construction, based on plane A, is shown in Fig. 2-1.

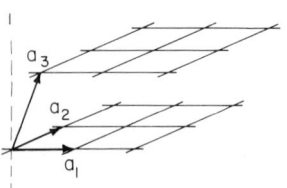

Fig. 2-1 Perpendicular rotation axis (1): Trigonal system for planes A,B,C,D,E.

2. *Monoclinic System.* A perpendicular rotation axis of order 2 on plane A gives rise to two possible lattices. If planes A are stacked directly above each other, a_3 is normal to the plane (Fig. 2-2). If the axis goes alternately through

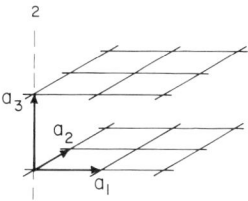

Fig. 2-2 Perpendicular rotation axis (2): monoclinic system for plane A. Orthorhombic system for planes B,C.

lattice points and cell midpoints of consecutive planes, a_3 points in the direction of displacement of adjacent planes (Fig. 2-3). Both lattices are described by the generating element 2 and a reflection plane m^{I} normal to the twofold axis.

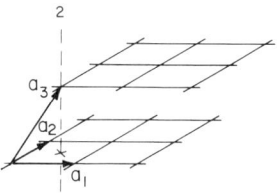

Fig. 2-3 Perpendicular rotation axis (2) with alternate stacking: monoclinic system for plane A, orthorhombic system for planes B,C.

3. *Orthorhombic System.* A twofold normal axis combined with either planes B or C leads to the existence of three perpendicular twofold axes, and the possibility of parallel or alternating displacement of planes, as already discussed, leads to four distinct lattice types. All are fully described by a set of three mutually perpendicular reflection planes $m^{\mathsf{I}}, m^{\mathsf{-}}, m^{\mathsf{\prime}}$. The superscripts are designed to distinguish between the three planes by labeling their orientation with respect to the major rotation axes (see Appendix 1 for detailed definitions).

The construction of these four lattices is similar to that of Figs. 2-2 and 2-3 with the additional conditions on the basis vectors $\mathbf{a}_1$ and $\mathbf{a}_2$ specified for planes B or C.

4. *Rhombohedral System.* A threefold normal axis can be constructed from a stacking of planes E. This axis goes through the lattice points of every third plane. For the two intervening planes the threefold axis passes through the center of the equilateral triangles of the planar cell. From the first to the second plane the cells are turned through 180°. This creates a single lattice, shown in Fig. 2-4, with the generating elements $\bar{1}, 3, 2\,\bar{}$.

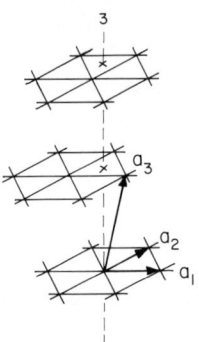

Fig. 2-4 Perpendicular rotation axis (3): rhombohedral system for triple stacking of plane E.

5. *Tetragonal System.* A fourfold normal axis on plane D, which may go through lattice points or square centers, gives rise to two lattices, both specified by the generating elements $\bar{1}, 4, 2\,\bar{}$. The lattices are shown in Figs. 2-2 and 2-3, with $a_1 = a_2$, and their included angle 90°. Here the twofold axis becomes fourfold.

6. *Hexagonal System.* A sixfold normal axis is obtained on plane E by parallel plane displacement. It produces one lattice (Fig. 2-5) with generating elements $\bar{1}, 6', 2\,\bar{}$.

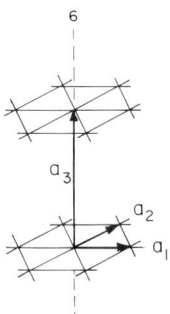

Fig. 2-5 Perpendicular rotation axis (6): hexagonal system for simple stacking of plane E.

7. *Cubic System.* If the orthorhombic system's three normal twofold axes are of equal length, additional symmetries characteristic of a cube are allowed. The four lattices of the orthorhombic system degenerate into three distinct cubic lattices: simple, face-centered, or body-centered. The simple cubic lattice is given in Fig. 2-2, with $a_1 = a_2 = a_3 = a$, and all three basis vectors orthogonal. The body-centered lattice is generated from Fig. 2-3 by taking $a_1 = a_2 = a$ and at right angles, and $a_3 = \frac{\sqrt{3}}{2} a$ at an angle $\tan^{-1}(1/\sqrt{2}) = 35.3°$ with the plane. The characteristic cube has dimension a. Finally, the face-centered lattice derives from Fig. 2-3 by taking $a_1 = a_2 = a$ and at right angles, but with $a_3 = a$ at an angle of 45° with the plane. Here the basic cube has dimension $\sqrt{2}a$. For all three lattices, the generating elements can be taken as 3 (along the cube body diagonal), 4 (normal to a cube face), and $\bar{1}$.

The foregoing list comprises 14 distinct lattices. For macroscopic purposes, however, the 14 lattices are described by the seven symmetry groups that characterize the seven crystal systems.

The seven symmetry groups are constructed from their generating elements. Once the seven groups are known it is an easy matter to derive all three-dimensional point groups by using the method applied in Section 2-4. We need only tabulate all their distinct subgroups. The totality of groups and subgroups of these symmetry groups yields 32 distinct point groups, which are indeed the point groups of three-dimensional space.

Appendix 1 lists both the complete specification of symmetry elements in three dimensions and the 32 point groups and their symmetry elements. The different groups are associated with the crystal system of the least symmetric lattices in which they first arise although, just as in the two-dimensional case, lattices in a system of higher symmetry also give rise to subgroups of such low symmetry.

We have derived the totality of distinct macroscopic symmetries by a geometrical construction of possible lattice types, followed by a mathematical deduction of the symmetry groups and their subgroups associated with such lattices. For our purposes, no additional characterization of a crystal on an atomic level (crystallography) is needed, and we can now focus directly on the consequences of endowing solids with any one of the allowed point group symmetries.

Problems

2.1. List the complete set of symmetry operations in the plane of an equilateral triangle. Show that this set contains as a subset the symmetry operations of an isosceles triangle.

2.2. Construct the multiplication table for the square lattice of type D.

2.3. Draw a geometrical pattern at each lattice site of the two-dimensional lattice C which leads to (a) the two-dimensional group $(2) = 1, 2$; (b) the two-dimensional group $(m) = 1, m$.

2.4. Verify the list of point groups in two dimensions given by Table 2.3.

2.5. Show that all the groups and subgroups based on the plane lattices B and C have a one-to-one correspondence and are formally identical.

2.6. Show that a rotation-inversion operation can also be described by a rotation-reflection.

2.7. Write out the multiplication table for the rhombohedral group (32) using geometrical construction of the symmetry operations.

2.8. Determine all the distinct subgroups of the rhombohedral lattice symmetry, and classify them according to the order of their rotation axis of highest symmetry.

2.9. Show that the generating elements indicated in the group label $(4/m)$ suffice to generate the whole group. Similarly for the group $(\bar{3}m)$.

2.10. Which of the seven crystal systems in three dimensions can lead to the crystal symmetry described by the group $(2/m) = 1, 2, \bar{1}, m'$?

2.11. The orthorhombic group (mmm) contains three mirror planes. Are these mirrors compatible with the three mirror planes of the group $(3m)$? Explain.

2.12. Specify a basis of geometrical design located on the lattice points of the hexagonal lattice which gives the crystal the symmetry (6) or $(\bar{6})$.

2.13. Does a basis of fourfold symmetry, placed on the tetragonal lattice, always give rise to a fourfold axis of the crystal? Explain.

2.14. Show that the construction of the body- and face-centered lattices specified in Section 2-5 leads to the familiar cube with lattice points at the corners, and at the cube center or the cube faces, respectively.

2.15. Verify by explicitly carrying out the covering operations in the group (42) that the twofold axes fall into three different classes. (See Appendix 1 for the definition of class.)

Bibliography

J. F. Nye, *Physical Properties of Crystals,* Oxford Univ. Press, London and New York (1957), Appendix B.

C. Kittel, *Introduction to Solid State Physics,* 3rd ed., Wiley, New York (1966), Chapter 1.

M. J. Buerger, *Elementary Crystallography*, Wiley, New York (1956).

M. A. Jaswon, *Introduction to Mathematical Crystallography*, American Elsevier, New York (1965).

R. S. Knox and A. Gold, *Symmetry in the Solid State*, W. A. Benjamin, Inc., New York (1964), Chapters 9 and 12.

CHAPTER 3

Mathematical Description of Crystal Properties

The physical properties of a crystal describe its response to externally applied influences such as forces and fields. In this description *directions* play an important role. Most of the applied forces and the responses of the crystal are represented by vectors or other quantities that are direction dependent. Furthermore, the nature of the response of the crystal is governed by its built-in directions of symmetry. All such directional aspects are readily included in a mathematical formulation that is based on components of directed quantities resolved in a coordinate system tied to the crystal.

3.1. THE SYMMETRY OF PHYSICAL PROPERTIES

Let us consider a crystal fixed in space, its properties described with respect to the coordinate system O. If the crystal possesses point symmetry, there exist a number of other systems $O', O'', \ldots$, one for each symmetry operation, in which the description of the physical properties must have identical form. Crystal symmetry makes these coordinate systems indistinguishable. Therefore, *a proper description of the physical properties of a crystal is invariant under the operations of its symmetry group.*

In order to assure that this required invariance is indeed maintained in the formal description of any particular given interaction, we must establish how the description itself transforms under the operations relating different coordinate systems. In general, this transformation depends very specifically on the physics of the interaction in question. But for the large class of interactions describable by *tensors* the transformation has a common formulation. Furthermore, tensor transformation is closely related to that of the coordinates themselves and makes use of the same basic quantities, so that this whole class of interactions can be treated without requiring new or specialized mathematics.

In restricting our treatment to tensor interactions, we nevertheless cover a very wide range of physical properties of interest. In fact, most phenomena have at least some aspects that can be described by tensors in the sense of a first approximation. For example, the dielectric constant is a tensor, while dielectric breakdown needs a more complex description. Similarly, the elastic constants are the tensors of reversible elasticity, but plastic deformation occurring at large strains is not so describable. Of course, the mathematical description of everyone of these phenomena must be in accord with crystal symmetry.

3.2. RESTRICTIONS ON MATTER TENSORS

If the fields and currents interacting with a crystal are related to vectors and tensors, the response of the crystal gives rise to new but similar vectors and tensors. Simple examples of such interactions are

$$\text{Electric polarization} \quad \mathbf{P} = \epsilon_0 \alpha \mathbf{E} \quad (3\text{-}1\text{a})$$

$$\text{Hall effect} \quad \mathbf{E}_H = R \mathbf{J} \times \mathbf{B} \quad (3\text{-}1\text{b})$$

The matter constant α is the *electric susceptibility* giving a linear connection between the vectors $\mathbf{P}$ and $\mathbf{E}$. The Hall effect relates a pair of vectors ($\mathbf{J}$ and $\mathbf{B}$) to a third ($\mathbf{E}_H$) through the *Hall constant R*. In general, α or R is a function of the directions of all the vectors involved. This function, however, is not arbitrary.

First of all, it is restricted by the purely formal nature of the interaction. Thus, in Eq. (3-1a), α connects two vectors, each described by three independent components. Written out in terms of components in a given reference system, Eq. (3-1a) takes the form

$$P_i = \epsilon_0 \sum_j \alpha_{ij} E_j, \quad i,j = 1, 2, 3. \quad (3\text{-}2)$$

It leads to nine independent and constant components of α_{ij} in the chosen reference system, no matter what the magnitudes and directions of $\mathbf{P}$ and $\mathbf{E}$. Hence, the same number of independent components must suffice in all reference systems. A similar argument applied to Eq (3-1b) limits R to a maximum of 27 independent constants.

Additional restrictions arise from the physical symmetries inherent in the interaction considered. Inherent symmetries primarily describe the behavior of the physical phenomenon with respect to space or time inversion. Time inversion symmetry manifests itself in the principles of thermodynamics, such as conservation of energy, or in the restrictions on transport phenomena given by the Onsager relations. It imposes, for example, the restriction $\alpha_{ij} = \alpha_{ji}$. Space inversion characteristics are related to a particular choice of fundamental entities in a theory, such as the designation of charge as a true scalar in electromagnetism, or they result from the intrinsic behavior of the interaction under parity. In our discussion inherent symmetries of this nature are either assumed or proved in sketchy form.

Finally, a set of matter constants such as α_{ij} in Eq. (3-2) is restricted by crystal symmetry proper. The description of the physical properties of the crystal must have the identical form in all the coordinate systems connected by point symmetry operations. This condition of invariance of the physical properties under crystal symmetry imposes additional conditions on the α_{ij}, and generally leads to a reduction in the number of independent components by requiring that entries for some (ij) vanish and that there be relations between the remaining ones.

In principle, anisotropic physics can be developed without explicit introduction of crystal symmetry. But in practice the simplification and reduction occurring when all known conditions of symmetry are considered from the beginning is of such importance that it deserves detailed exposition. This is the effect of crystal symmetry with which we are primarily concerned.

The program for ensuring the invariance of the description of physical properties under crystal symmetry is carried out in two stages. First, it is necessary to know the transformation properties of the physical quantities in question. This requires treating the transformations of tensors. Second, we carry out explicitly the transformation of the matter constants under all applicable symmetry operations, and then require that each component maintain the same value under all such transformations. This requirement is expressed by a set of linear relations between the various tensor components. Generally, their solution leads to an appreciable reduction in the number of nonvanishing tensor components: *the tensor exhibits crystal symmetry.*

3.3. ORTHOGONAL TRANSFORMATIONS

Let us choose an orthogonal system of axes for the coordinate system O (not necessarily identical with crystal axes of symmetry); O is described by three orthonormal basis vectors $\mathbf{e}_i$

$$\mathbf{e}_i \cdot \mathbf{e}_j = \delta_{ij} \tag{3-3a}$$

$$\mathbf{e}_i \times \mathbf{e}_j = \epsilon_{ijk} \mathbf{e}_k \quad \text{(if right-handed)} \tag{3-3b}$$

where δ_{ij} is the Kronecker delta ($= 1$ for $i = j$, $= 0$ for $i \neq j$), and ϵ_{ijk} is the antisymmetric triple product ($= 1$ for i, j, k in cyclic order; -1 for i, j, k in reverse order; and $= 0$ if two or more indices are the same).

3.3. Mathematical Description of Crystal Properties

Another orthogonal system O', with a set of basis vectors e_i' rotated with respect to O is related linearly to the first

$$e_i' = \sum_j R_{ij} e_j, \qquad i = 1, 2, 3 \tag{3-4}$$

The coefficients R_{ij} are the direction cosines between the old and new vectors

$$R_{ij} = e_i' \cdot e_j \tag{3-5}$$

A few well-known properties of the transformation parameters R_{ij} are

orthogonality $\qquad\qquad \sum_i R_{ij} R_{ik} = \delta_{jk} \tag{3-6a}$

normalization $\qquad\qquad \sum_i R_{ji} R_{ki} = \delta_{jk} \tag{3-6b}$

inverse transformation $\qquad R_{ij}^{-1} = R_{ji} \tag{3-7}$

The nine coefficients R_{ij} can be written as a square array

$$(R_{ij}) = \begin{pmatrix} R_{11} & R_{12} & R_{13} \\ R_{21} & R_{22} & R_{23} \\ R_{31} & R_{32} & R_{33} \end{pmatrix} \tag{3-8}$$

and if we represent the basis vectors by columns and use the rules of matrix multiplication, Eq. (3-4) can also be expressed in the form

$$\begin{pmatrix} e_1' \\ e_2' \\ e_3' \end{pmatrix} = \begin{pmatrix} R_{11} & R_{12} & R_{13} \\ R_{21} & R_{22} & R_{23} \\ R_{31} & R_{32} & R_{33} \end{pmatrix} \begin{pmatrix} e_1 \\ e_2 \\ e_3 \end{pmatrix} \tag{3-9}$$

Just like the basic vectors themselves, the rows and columns of Eq. (3-8) can be considered orthogonal vectors of unit length. These properties follow from Eq. (3-6). Consequently, only three of the nine entries in Eq. (3-8) are truly independent, as is expected since the coordinate systems O and O' are connected by three independent rotations and their three direction cosines as given by Eq. (3-5).

In addition to defining the rotation from O to O', (R_{ij}) also contains information about the sense of the new coordinate system. It is easy to show that if the determinant $|R_{ij}|$ has the value $+1$, O and O' have the same sense. For example, both are right-handed. On the other hand, if $|R_{ij}| = -1$, the sense of O' is opposite to that of O. If O is left-handed, then O' is right-handed, and conversely. In this case the transformation (R_{ij}) includes an inversion or an improper rotation.

So far we have been concerned with the transformation of the basis vectors from O to O'. We can now ask how this transformation affects the description of a fixed point in the crystal. If the point has the coordinates (c_1, c_2, c_3) in O it is given by the vector **v**

$$\mathbf{v} = \sum_k c_k \mathbf{e}_k \tag{3-10}$$

In O' the *same* vector is given by new coordinates (c_1', c_2', c_3')

$$\mathbf{v} = \sum_j c_j' \mathbf{e}_j' \tag{3-11}$$

Equating the scalar product of $\mathbf{e}_i'$ with **v** in Eq. (3-11) to the scalar product of $\mathbf{e}_i'$ with **v** in Eq. (3-10), we obtain

$$\sum_j \mathbf{e}_i' \cdot \mathbf{e}_j' c_j' = \sum_k \mathbf{e}_i' \cdot \mathbf{e}_k c_k \tag{3-12}$$

and simplifying through use of Eqs. (3-3a) and (3-5), we arrive at the final result

$$c_i' = \sum_k R_{ik} c_k \tag{3-13}$$

Hence we have shown that *the coordinates of a point transform exactly like basis vectors.*

3.4. CONSECUTIVE TRANSFORMATIONS AND MATRIX REPRESENTATIONS OF THE SYMMETRY GROUP

In Section 3-3 the transformation (R_{ij}) took us from O to O'. If there is a second transformation (R'_{ij}) taking O' into O'', we write,

$$\mathbf{e}_i'' = \sum_j R'_{ij} \mathbf{e}_j' \tag{3-14}$$

We want to find the single transformation taking O into O''. Using Eq. (3-4) we eliminate the basis vectors $\mathbf{e}_j'$ in Eq. (3-14) and obtain (after relabeling the dummy indices which are summed)

$$\mathbf{e}_i'' = \sum_k \left(\sum_j R'_{ij} R_{jk} \right) \mathbf{e}_k \tag{3-15}$$

Hence if (R''_{ik}) is the transformation from O to O'', Eq. (3-15) gives the rule of combination

$$R''_{ik} = \sum_j R'_{ij} R_{jk} \tag{3-16}$$

This is the rule of *matrix multiplication*. Therefore the square array (R_{ij}) introduced in Eq. (3-8) is identified as a matrix and it follows all the well-known rules of combinations of these entities.

So far we have dealt with transformations between any two arbitrary coordinate systems. Now suppose that we have a crystal of given symmetry described by a set of symmetry elements or symmetry operations. Each symmetry operation has a corresponding coordinate transformation, and from the discussion of this section we can also construct the transformation describing two consecutive symmetry operations. Thus, if the element X leads to transformation (X_{ij}) and Y leads to (Y_{ij}), then the consecutive application YX is given by the transformation

$$(YX)_{ik} = \sum_j Y_{ij} X_{jk} \tag{3-17}$$

It is clear that the set of elements describing the symmetry group of the crystal generates a corresponding set of transformation matrices, and under the rule of combination given by Eq. (3-16) the matrices follow the same multiplication table, as for example in Eq. (2-2), as the abstract symmetry elements. Such a set of matrices is said to form a *matrix representation* of the abstract group. The matrices describing the symmetry transformations of the coordinate system are a three-dimensional matrix representation of the symmetry group. The same group has representations of other dimensions, depending on the particular physical property the representation is to describe.

3.5. TRANSFORMATION PROPERTIES OF TENSORS

Equation (3-2) is an example of a linear relationship between two vectors. The electric susceptibility α_{ij} describes the polarization **P** in given field **E**. In a new coordinate system O' this equation must relate vector components with respect to O', and hence the set α_{ij} will also take on a new form. Since we know from Eq. (3-13) how vector components (coordinates) transform under (R_{ij}), the transformation of α_{ij} can be constructed directly by rewriting the vectors in Eq. (3-2) in terms of their new components. Upon application of the orthogonality relations of Eq. (3-6) the equation can be brought back into the original form (but with a set α'_{ij} replacing α_{ij}) and given by

$$\alpha'_{ij} = \sum_{k,l} R_{ik} R_{jl} \alpha_{kl} \tag{3-18}$$

The rule of transformation of Eq. (3-18) defines (α_{ij}) as a *second rank tensor*. The tensor components are characterized by two indices (each running from 1 to 3) and they transform under a coordinate transformation (R_{ij}) like products of coordinates.

For our purposes, a tensor of rank m is a quantity T with m indices, transforming under a coordinate transformation (R_{ij}) like

$$T'_{abc\cdots m} = \sum_{\alpha,\beta,\gamma,\cdots,\nu} R_{a\alpha} R_{b\beta} R_{c\gamma} \cdots R_{m\nu} T_{\alpha\beta\gamma\cdots\nu} \qquad (3\text{-}19)$$

This definition includes scalars (zero rank) and vectors (first rank), and classifies all higher tensors, such as α_{ij} (second rank), Hall constant R_{ijk} (third rank), and so on.

With respect to our applications of tensor transformation, two additional observations are of importance.

In general, it is important to specify whether tensor components transform *contravariantly* (like coordinate differences) or *covariantly* (like the gradient of a scalar function) with respect to each one of their indices. This distinction disappears for transformations connecting rectilinear orthogonal systems, and does not play any role in the applications to follow.

Vectors denoting physical quantities show both possible kinds of behavior under inversion (parity). A true vector, such as the electric field **E** is a *polar vector*. Upon inversion of the coordinate system, all components of **E** change sign. On the other hand, vectors such as the magnetic field do not change sign under inversion. They are labeled *axial vectors*. As a consequence, tensor indices referring to axial vector components will not transform strictly according to Eq. (3-19) under transformations (R_{ij}) that define improper rotations (i.e., that include an inversion). There are two ways of handling this problem. Either we specify for each index whether it refers to a polar or an axial vector. Or, since the net transformation will depend only on whether the tensor has an even or odd number of axial indices, we can label the tensor itself as polar or axial. The second approach is most convenient. We therefore define two types of tensors. A *polar tensor* transforms strictly like a product of coordinates under either rotations or rotation-inversions. An *axial tensor* transforms like a product of coordinates under rotations, but like (-1) times the product of coordinates under rotation-inversions. In this manner, we will not have to assign polar or axial transformation properties to specific tensor indices.

3.6. A PRACTICAL RULE FOR CARRYING OUT TENSOR TRANSFORMATIONS

Although Eq. (3-19) gives a perfectly general rule for carrying out the transformation of tensor components for a given (R_{ij}), in practice it is often convenient to supplement this formal rule by a more direct approach. This new rule reduces greatly the amount of manipulation of the transformation matrices, and thus also helps in minimizing errors that often creep into the elaborate bookkeeping of indices and entries demanded by Eq. (3-19).

The practical rule is based on the fact that the transformation of Eq. (3-19) is the same as that of a product of coordinates. Thus, if x_i stands for one of the components of the coordinate triple (x, y, z), it obeys the transformation rule $x_i' = \Sigma_j R_{ij} x_j$. Therefore, Eq. (3-19) holds equally true for the m-tuple coordinate product

$$(x_a')(x_b')(x_c') \cdots (x_m')$$

and we can say that the tensor component $T'_{abc\cdots m}$ transforms like its corresponding coordinate product. Hence, if we identify each index of the tensor component with a coordinate triple (x, y, z), we can carry out explicitly the transformation of the tensor in terms of the transformation of the corresponding coordinate product. This identification is very easy. For example, the component α_{12} transforms like the coordinate product $(x)(y)$, α_{32} like $(z)(y)$, or α_{33} like $(z)(z)$. Once we know the coordinate product involved, it is equally easy to carry out the tensor transformation, because the transformation of each coordinate is given explicitly in terms of the R_{ij}. This is best demonstrated by treating some simple examples.

Given the transformation of coordinates describing a rotation about the z axis

$$R_{ij} = \begin{pmatrix} \cos\theta & \sin\theta & 0 \\ -\sin\theta & \cos\theta & 0 \\ 0 & 0 & 1 \end{pmatrix} \qquad (3\text{-}20)$$

The coordinates follow the transformation

$$x' = x\cos\theta + y\sin\theta, \quad y' = -x\sin\theta + y\cos\theta, \quad z' = z$$

The tensor component α'_{12} transforms like $(x')(y')$ and can therefore be written

$$\begin{aligned}
\alpha'_{12} \sim (x')(y') &= (x\cos\theta + y\sin\theta)(-x\sin\theta + y\cos\theta) \\
&= -(x)(x)\cos\theta\sin\theta - (y)(x)\sin^2\theta + (x)(y)\cos^2\theta \\
&\quad + (y)(y)\sin\theta\cos\theta \\
&\sim -\alpha_{11}\cos\theta\sin\theta - \alpha_{21}\sin^2\theta + \alpha_{12}\cos^2\theta + \alpha_{22}\sin\theta\cos\theta
\end{aligned}$$

or

$$\alpha'_{12} = (-\alpha_{11} + \alpha_{22})\sin\theta\cos\theta - \alpha_{21}\sin^2\theta + \alpha_{12}\cos^2\theta$$

Similarly,

$$\alpha'_{32} \sim (z')(y') = (z)(-x \sin\theta + y \cos\theta)$$

or

$$\alpha'_{32} = -\alpha_{31} \sin\theta + \alpha_{32} \cos\theta$$

This rule, which applies to arbitrary transformations, simplifies greatly the problem of computing transformed tensor components. Formally, it amounts to inverting the usual procedure. Rather than follow the formal rule of first compounding the transformation products before summing terms, we multiply directly the individual coordinate transformations to obtain the final products. Note that in general the *order* of the coordinates in the factors must be retained.

Practically, this method has the added advantage that it generates immediately the tensor components of interest without our having to go through the construction of the entire set of composite transformation matrices demanded by Eq. (3-19).

3.7. DETERMINATION OF THE INDEPENDENT CONSTANTS

It is obvious how the foregoing results can be applied to determine the form of a tensor in a crystal of specified symmetry. Let us start by defining the components of a tensor, for instance T_{abc}, in one coordinate system. Given a symmetry operation R, we then construct the corresponding coordinate transformation (R_{ij}), and find the transformed tensor components T'_{abc}. Since the two coordinate systems are indistinguishable, we require $T'_{abc} = T_{abc}$ for all components. This yields relations between the original tensor components. Additional relations follow by applying all symmetry operations, one after another, to each tensor component and requiring that it go into itself after the transformation.

Once all relations between the tensor components are known, we solve the whole set of equations and determine those components which must vanish and any relations between the nonvanishing terms.

This is a tedious process. It can be simplified in practice in a number of ways: (1) It is not necessary to consider all symmetry operations, but only the generating elements of the symmetry group. (2) If the generating elements have entries only of value ± 1 (i.e., in a special coordinate system), the transformation of the tensor components is simplified because each row or column of the transformation matrices has only a single entry. Consequently, each tensor component transforms into a single new component.

A convenient set of generating elements and of coordinate systems in which they have the desired form is listed in Appendix 2 for all point groups. With this information, it is a straightforward process to construct the form of a tensor

3.7 Mathematical Description of Crystal Properties

compatible with crystal symmetry. It must be kept in mind, though, that the form of the tensor so obtained holds only in the special coordinate system in which the generating elements are defined. This coordinate system may not always coincide with the system of crystal axes.

To illustrate the method, let us construct the general second-order tensor in the group (4). Appendix 2 lists the generating element as

$$4 = \begin{pmatrix} 0 & 1 & 0 \\ -1 & 0 & 0 \\ 0 & 0 & 1 \end{pmatrix}$$

Application of this transformation yields the following set of relations.

$$\begin{aligned}
\alpha'_{11} &\sim (x')(x') = (y)(y) &&= \alpha_{22}, &&\text{or} &&\alpha_{22} = \alpha_{11} \\
\alpha'_{12} &\sim (x')(y') = (y)(-x) &&= -\alpha_{21}, &&&&-\alpha_{21} = \alpha_{12} \\
\alpha'_{13} &\sim (x')(z') = (y)(z) &&= \alpha_{23}, &&&&\alpha_{23} = \alpha_{13} \\
\alpha'_{21} &\sim (y')(x') = (-x)(y) &&= -\alpha_{12}, &&&&-\alpha_{12} = \alpha_{21} \\
\alpha'_{22} &\sim (y')(y') = (-x)(-x) &&= \alpha_{11}, &&&&\alpha_{11} = \alpha_{22} \\
\alpha'_{23} &\sim (y')(z') = (-x)(z) &&= -\alpha_{13}, &&&&-\alpha_{13} = \alpha_{23} \\
\alpha'_{31} &\sim (z')(x') = (z)(y) &&= \alpha_{32}, &&&&\alpha_{32} = \alpha_{31} \\
\alpha'_{32} &\sim (z')(y') = (z)(-x) &&= -\alpha_{31}, &&&&-\alpha_{31} = \alpha_{32} \\
\alpha'_{33} &\sim (z')(z') = (z)(z) &&= \alpha_{33}, &&&&\alpha_{33} = \alpha_{33}
\end{aligned}$$

These nine relations require that four components vanish, and that in addition there be two connections between the remaining five. Hence the most general second rank tensor in the system (4) has the form

$$\alpha_{ij} = \begin{pmatrix} \alpha_{11} & \alpha_{12} & 0 \\ -\alpha_{12} & \alpha_{11} & 0 \\ 0 & 0 & \alpha_{33} \end{pmatrix}$$

with three independent constants.

If we further require that the tensor α_{ij} be symmetric, it is reduced to diagonal form. Since $\alpha_{11} = \alpha_{22}$, the crystal system (4) behaves with respect to symmetric second rank tensors for all rotations in the base plane like an isotropic material.

This method for constructing tensors in symmetric systems is named the *direct inspection method*, because it leads to the answer practically by inspection only.

We can use it readily to obtain even rather high rank tensors. Most common tensors have, of course, already been constructed and are given in many texts. There the method gives a convenient check on the consistency of the component schemes. In all cases, however, its application requires that we also know whether the tensor in question is polar or axial. The problems at the end of this chapter list a variety of applications that bring out additional details of the method.

In addition to the 32 groups, we are often also interested in the tensor symmetry to be expected in an isotropic body or a material with axial symmetry. This symmetry cannot be deduced using the present scheme. It is treated in Chapter 4.

Problems

3.1. If we ignore the restrictions imposed by the vector product $\mathbf{J} \times \mathbf{B}$, Eq. (3-1b) has the most general form (in terms of components)

$$E_i = \sum_{j,k} R_{ijk} J_j B_k$$

Show that

(a) R_{ijk} has a maximum of 27 terms;

(b) under the *intrinsic* symmetry $R_{ijk} = -R_{jik}$, the number of independent terms is reduced to 9;

(c) under the additional *crystal* symmetry that R_{ijk} is nonvanishing only if k differs from both i and j, the number of independent terms is 3. How many entries are nonzero?

3.2 Prove Eq. (3-6a) by explicitly requiring Eq. (3-3a) to hold for the new basis vectors $\mathbf{e}_i'$.

3.3 If R describes the transformation $\mathbf{e}_i' = \sum R_{ij} \mathbf{e}_j$, the inverse transformation R^{-1} is $\mathbf{e}_i = \sum R_{ij}^{-1} \mathbf{e}_j'$. Show that Eq. (3-7) is a sufficient condition for recovering the original set of basis vectors when R^{-1} is applied to $\mathbf{e}_i'$. Is it also a necessary condition?

3.4. Construct the rotation matrix R_{ij} such that $\mathbf{e}_1'$ makes equal angles with the $x, y,$ and z axes, and $\mathbf{e}_2'$ is in the xy plane.

3.5. Given a transformation R_{ij} with the diagonal entries $R_{11} = a, R_{22} = b, R_{33} = c$, with $1 \geqslant |a| \geqslant |b| \geqslant |c|$.
(a) Show that if $1 + c^2 \geqslant a^2 + b^2$, the *normalization* conditions of Eqs. (3-6a) and (3-6b) allow a solution with the symmetry $R_{ij}^2 = R_{ji}^2$.
(b) Show further that the *orthogonality* conditions of the same equations are satisfied only if one or more of the eight relations

$$(\pm a \pm b \pm c)^2 = 1$$

are obeyed, and construct the full matrix for $a = 0.9, b = 0.6, c = 0.5$.

Mathematical Description of Crystal Properties

3.6. The determinant $|R_{ij}|$ is given by

$$|R_{ij}| = \sum_{l,m,n} \epsilon_{ijk} R_{il} R_{jm} R_{kn}$$

Use Eq. (3-3b) to show that if $|R_{ij}| = \pm 1$ for Eq. (3-4), then
$e'_i \times e'_j = \pm \epsilon_{ijk} e'_k$.

3.7. Verify that the set of matrices

$$\begin{pmatrix} 1 & 0 \\ 0 & 1 \end{pmatrix}, \begin{pmatrix} -1 & 0 \\ 0 & -1 \end{pmatrix}, \begin{pmatrix} 1 & 0 \\ 0 & -1 \end{pmatrix}, \begin{pmatrix} -1 & 0 \\ 0 & 1 \end{pmatrix}$$

follows the same multiplication table of Eq. (2-2) if the matrices corres=
pond to the elements $1, 2, m, m'$.

3.8 Prove Eq. (3-18), and show that the transformation can also be written

$$\alpha'_{ij} = \sum_{k,l} R_{ik} \alpha_{kl} R^{-1}_{lj}$$

3.9. Show that for transformations between rectilinear orthogonal coordinate systems, the coordinates (x_1, x_2, x_3) and the gradient components $(\partial\phi/\partial x_1, \partial\phi/\partial x_2, \partial\phi/\partial x_3)$ transform according to the same rule.

3.10. Show that the vector product $\mathbf{A} \times \mathbf{B}$ of two polar vectors is an axial vector, and that the triple product $\mathbf{A} \times \mathbf{B} \cdot \mathbf{C}$ is a scalar only if $\mathbf{C}$ is axial. If $\mathbf{C}$ is polar, the triple product is an axial scalar (pseudoscalar).

3.11. Given that the magnetic induction $\mathbf{B}$ is an axial vector, determine whether the Hall constant tensor R_{ijk} of Problem 3-1 is axial or polar.

3.12. Given the transformation

$$\mathcal{R}_{ij} = \begin{pmatrix} \frac{1}{\sqrt{3}} & \frac{1}{\sqrt{2}} & -\frac{1}{\sqrt{6}} \\ \frac{1}{\sqrt{3}} & 0 & \frac{2}{\sqrt{6}} \\ -\frac{1}{\sqrt{3}} & \frac{1}{\sqrt{2}} & \frac{1}{\sqrt{6}} \end{pmatrix}$$

If α_{ij} is a polar tensor, use the rules of Section 3-6 to find the transformed component α'_{23}. Is this result changed if α_{ij} is an axial tensor?

3.13. The third rank tensor T_{ijk} is totally symmetric in its indices. Show that under the transformation R_{ij} of Problem 3-12

$$T'_{222} = \frac{1}{3\sqrt{3}} T_{111} + \frac{2}{\sqrt{6}} T_{113} + \frac{2}{\sqrt{3}} T_{133} + \frac{4}{3\sqrt{6}} T_{333}$$

3.14. Find the form of the general second rank tensor for crystals of symmetry (mm) and (32). Treat the cases for both polar and axial tensors.

3-15. In a certain monoclinic crystal, the dielectric constant tensor has the value

$$\epsilon_{ij} = \begin{pmatrix} 5 & 2 & 0 \\ 2 & 2 & 0 \\ 0 & 0 & 3 \end{pmatrix}$$

with respect to the chosen crystal axes. Find the transformation to a new coordinate system in which ϵ'_{ij} is a diagonal tensor, and determine the principal dielectric constants ϵ'_{ii}.

3-16. Use the direct inspection method to show that in the tetragonal system $(\bar{4}2m)$ a polar third rank tensor has the nonvanishing components

$$T_{123} = T_{213}, \quad T_{132} = T_{231}, \quad T_{312} = T_{321}$$

Bibliography

J. F. Nye, *Physical Properties of Crystals,* Oxford Univ. Press, London and New York (1957), Chapters 1 and 2.

W. P. Mason, *Crystal Physics of Interaction Processes,* Academic Press, New York (1966), Chapter 2.

S. Bhagavantam, *Crystal Symmetry and Physical Properties,* Academic Press, New York (1966), Chapters 1–3.

CHAPTER 4

Tensor Symmetry and Linear Vector Spaces

Most practical problems involving the commonly used tensors in symmetric systems can be handled using the techniques of Chapter 3. However, there are special aspects of tensors, such as the effect of symmetry groups with an infinite number of elements, or the structure and construction of tensors of very high order, that are best treated using a more general approach based on concepts of group theory. This chapter gives an introduction to these concepts as they apply to the problems of interest in crystal physics. Far from being a complete exposition, it aims mainly at developing some additional practical tools for dealing with tensors, using only a minimum of the often rather abstract formal theory. However, detailed understanding of these methods is not a prerequisite for following most of the material of the subsequent chapters. The subject can be taken up whenever a concrete need for it arises in specific applications.

4.1 TENSOR INVARIANTS

The components of a matter tensor T, defined with respect to a coordinate system, generally take on new values if the frame of reference for T is changed. However, there may exist some linear combinations of the tensor components that remain constant under all transformations. These combinations form a set of *invariants* of the tensor.

The invariants of a tensor are easily constructed by extending the methods introduced in Chapter 3 for handling the transformation of tensor components. According to Eq. (3-19) a tensor component transforms like a product of coordinates or, equivalently, like a product of vector components. This rule requires that we assign to each index of the tensor a different vector **V**, and then form the product by using the component of each **V** indicated by the particular value (1, 2, or 3) of the subscript labeling the tensor index. For example, the third rank tensor (T_{ijk}) has three indices, to which we assign in order the vectors $\mathbf{V}_1, \mathbf{V}_2, \mathbf{V}_3$. Then the rule states that T_{213} transforms like $(V_1)_2(V_2)_1(V_3)_3$, T_{112} like $(V_1)_1(V_2)_1(V_3)_2$, and so on.

Hence we can identify each tensor component with its product of vector components. Further, the complete set of tensor components includes all possible combinations of products of vector components, and conversely, all the components of the tensor are obtained from the complete set of vector component products. Components of a tensor of nth rank and n-fold products of vector components, each vector occurring once, span the same space.

With this correspondence established, invariant linear combinations of tensor components are obtained simply by constructing those products of the vectors assigned to the tensor that remain constant under all transformations of the coordinate system. In other words, we search for the *scalars* which can be formed from the set of vectors $\mathbf{V}_1, \mathbf{V}_2, \mathbf{V}_3, \ldots$. For instance, the third rank tensor (T_{ijk}) has an invariant under rotation corresponding to the scalar triple product $\mathbf{V}_1 \times \mathbf{V}_2 \cdot \mathbf{V}_3$. Since no other combinations of the three vectors result in a scalar, this is the only invariant of (T_{ijk}). Written out, the triple product is

$$(V_1)_y(V_2)_z(V_3)_x - (V_1)_z(V_2)_y(V_3)_x + (V_1)_z(V_2)_x(V_3)_y$$

$$-(V_1)_x(V_2)_z(V_3)_y + (V_1)_x(V_2)_y(V_3)_z - (V_1)_y(V_2)_x(V_3)_z$$

and the corresponding tensor invariant is

$$T_{231} - T_{321} + T_{312} - T_{132} + T_{123} - T_{213} \tag{4-1}$$

Similarly, the invariant of a second rank tensor is given by the scalar product $\mathbf{V}_1 \cdot \mathbf{V}_2 = x_1 x_2 + y_1 y_2 + z_1 z_2$. (Here we use the shorthand notation for identifying vector components by their corresponding coordinates.) Hence the invariant of a second rank tensor (T_{ij}) is

$$T_{11} + T_{22} + T_{33} \tag{4-2}$$

When the same approach is applied to higher-order tensors that have more than one invariant it is necessary to choose, from among all possible scalars obtained by combining the given vectors, a *set of linearly independent invariants*. For instance, the four vectors $\mathbf{V}_1, \mathbf{V}_2, \mathbf{V}_3, \mathbf{V}_4$ can be combined to form the following six scalars.

$$(\mathbf{V}_1 \cdot \mathbf{V}_2)(\mathbf{V}_3 \cdot \mathbf{V}_4), \quad (\mathbf{V}_1 \cdot \mathbf{V}_3)(\mathbf{V}_2 \cdot \mathbf{V}_4), \quad (\mathbf{V}_1 \cdot \mathbf{V}_4)(\mathbf{V}_2 \cdot \mathbf{V}_3)$$
$$(\mathbf{V}_1 \times \mathbf{V}_2) \cdot (\mathbf{V}_3 \times \mathbf{V}_4), \quad (\mathbf{V}_1 \times \mathbf{V}_3) \cdot (\mathbf{V}_2 \times \mathbf{V}_4), \quad (\mathbf{V}_1 \times \mathbf{V}_4) \cdot (\mathbf{V}_2 \times \mathbf{V}_3) \tag{4-3}$$

However, since the entries in the second line are linear combinations of the scalars in the first line, there are only *three* independent scalars. Hence the fourth rank tensor (T_{ijkl}) has three invariants. They are not unique but can be formed as any three linearly independent combinations of the foregoing set of six scalars.

Strictly speaking, the formulation presented so far applies only to constructing linear combinations of tensor components invariant for arbitrary *rotations*. In order to include invariance under inversion, we must know whether the tensor in question is polar or axial. It is easily seen that all rotational invariants of even-rank polar tensors and odd-rank axial tensors remain invariant under inversion. Conversely, it follows that odd-rank polar and even-rank axial tensors have *no invariants* if inversion is included. In this case the invariants we have constructed for rotations are really pseudoscalars that reverse sign in a transformation of changing parity.

If a tensor has intrinsic symmetry, its invariants can be constructed by following the procedure for a general tensor and then introducing the additional symmetry afterward. However, this symmetry can also be taken into account right at the beginning. Thus, if two or more tensor indices are symmetric, we let their corresponding vectors be equal. For instance, if a fourth rank tensor has the intrinsic symmetry $T_{ijkl} = T_{jikl} = T_{ijlk}$, we let the vectors of Eq. (4-3) obey $\mathbf{V}_1 = \mathbf{V}_2$ and $\mathbf{V}_3 = \mathbf{V}_4$. It follows that the last two scalars in the first line become identical. Hence a tensor of this symmetry has only two invariants, as for instance

$$(V_1^2)(V_3^2), \quad (\mathbf{V}_1 \cdot \mathbf{V}_3)^2 \tag{4-4}$$

On the other hand, if two tensor indices are antisymmetric, we assign a single vector to the index pair, since the three antisymmetric combinations (*ij*) label the components of a vector (obtained by a vector product).

4.2. TENSOR SYMMETRY IN ISOTROPIC MATERIALS

In an isotropic medium all directions are equivalent. All tensors describing the properties of such a material must have a form that is independent of coordinate axes. The invariant expressions derived in Section 4-1 bear directly on the problem of constructing such tensors. Two results follow immediately. The number of independent tensor components is equal to the number of invariants. In addition, only the tensor components appearing in the invariants are nonvanishing.

To construct the explicit scheme of all tensor components using only the invariants is not always possible. But in all cases their use leads to great simplification. First of all, it is obvious that even though an invariant, such as given by Eq. (4-1), contains a large number of individual tensor components, these represent only a small number of basic types. For example, since all rotated coordinate systems are equivalent, the tensor components must certainly be invariant under an exchange of coordinate axes.

$$x \to y \to z \to x \tag{4-5}$$

It follows immediately that

$$T_{123} = T_{231} = T_{312} = -T_{321} = -T_{132} = -T_{213}$$

so that there is only one type of component in Eq. (4-1). It occurs in three entries of (T_{ijk}) as positive, in three other entries as negative, and all other entries are identically equal to zero. Similar reasoning demands that in Eq. (4-2) $T_{11} = T_{22} = T_{33}$, with all other entries vanishing.

This procedure applies equally well in the case of more than one invariant, as long as the number of invariants is equal to the number of types of components occurring in the invariants. If the number of component types is larger than the number of invariants, we must employ a somewhat more careful approach, because in that case not all component types are independent. To determine the relations between them we offer here a rule whose full justification will emerge in a later section (Section 4-3):

1. Construct from the initial set of invariants a new set of invariants orthogonal to each other.
2. Form all the additional linear combinations of tensor components, using the same component types, that are orthogonal to the invariant set.
3. The linear combinations obtained in step 2, not being invariants, must vanish. This condition establishes the required relations.

To apply the rule we must define orthogonality of two linear combinations of tensor components. It is a direct extension of the orthogonality of two vectors. *Two linear combinations of tensor components are orthogonal if the sum of the products of the coefficients of like components is equal to zero.*

As an example, let us apply this rule to obtain the isotropic tensor scheme of T_{ijkl}. According to Eq. (4-3), this tensor has three invariants, and if we choose the three scalars of the first line as the initial set, we obtain.

$$I_1 = T_{1111} + T_{2222} + T_{3333} + T_{1122} + T_{2211} + T_{1133} + T_{3311} + T_{2233} + T_{3322}$$

$$I_2 = T_{1111} + T_{2222} + T_{3333} + T_{1212} + T_{2121} + T_{1313} + T_{3131} + T_{2323} + T_{3232} \tag{4-6}$$

$$I_3 = T_{1111} + T_{2222} + T_{3333} + T_{1221} + T_{2112} + T_{1331} + T_{3113} + T_{2332} + T_{3223}$$

4.2 Tensor Symmetry and Linear Vector Spaces

These three invariants contain four types of components: $T_{1111}, T_{1122}, T_{1212}$, and T_{1221}. Since there are only three invariants, there must exist one relation between the nonvanishing components. The invariants in Eq. (4-6) are clearly not orthogonal. Following step 1 of the foregoing rule, we construct an equivalent orthogonal set

$$I'_1 = I_1 + I_2 + I_3, \quad I'_2 = 2I_1 - I_2 - I_3, \quad I'_3 = I_2 - I_3 \tag{4-7}$$

Since there are four component types, there exists one aditional linear combination using the same component types that is orthogonal to the three combinations of Eq. (4-7). This fourth linear combination is

$$-2(T_{1111} + T_{2222} + T_{3333}) + (T_{1122} + T_{2211} + T_{1133} + T_{3311} + T_{2233} + T_{3322})$$
$$+ (T_{1212} + T_{2121} + T_{1313} + T_{3131} + T_{2323} + T_{3232}) \tag{4-8}$$
$$+ (T_{1221} + T_{2112} + T_{1331} + T_{3113} + T_{2332} + T_{3223})$$

To find the relation of interest we note, first that all components of one type always occur as sums, so that they must all be equal to each other. Second, Eq. (4-8), not being an invariant, must vanish. Hence, the full scheme of the tensor (T_{ijkl}) in an isotropic material is given by

$$T_{1111} = T_{2222} = T_{3333}$$

$$T_{1122} = T_{2211} = T_{1133} = T_{3311} = T_{2233} = T_{3322}$$

$$\tag{4-9}$$

$$T_{1212} = T_{2121} = T_{1313} = T_{3131} = T_{2323} = T_{3232}$$

$$T_{1221} = T_{2112} = T_{1331} = T_{3113} = T_{2332} = T_{3223}$$

with

$$T_{1111} = T_{1122} + T_{1212} + T_{1221}$$

Thus, in isotropic matter (T_{ijkl}) has 21 nonvanishing components involving three independent constants.

The same procedure applies to constructing the tensor scheme in isotropic matter when the tensor has intrinsic symmetry, such as is implied in Eq. (4-4). In such cases, however, one has to be careful to define orthogonality because the contractions allowed by intrinsic symmetry will introduce additional numerical coefficients.

The construction of tensors in axially symmetric materials, that is, materials allowing arbitrary rotations about a single axis, follows very much the same principles. It will be exemplified by some problems at the end of this chapter.

4.3. INVARIANT SUBSPACES OF TENSOR COMPONENTS

The scalar invariants of Section 4-1 represent the linear combinations of tensor components derived from *scalar* combinations of the vectors representing each of the tensor indices. Obviously, these vectors can be combined also to form *vectors*. Just as the three components of a vector transform among themselves under coordinate transformations, the corresponding linear combinations of tensor components transform among themselves, and are said to form an *invariant subspace*.

The three linear combinations of the second rank tensor (T_{ij}) transforming like a vector correspond to the three components of the vector product $\mathbf{V}_1 \times \mathbf{V}_2 = (y_1 z_2 - z_1 y_2, z_1 x_2 - x_1 z_2, x_1 y_2 - y_1 x_2)$, and are

$$T_{23} - T_{32}, \quad T_{31} - T_{13}, \quad T_{12} - T_{21} \tag{4-10}$$

Equations (4-2) and (4-10) together define four linear combinations of tensor components forming one one-dimensional and one three-dimensional subspace. The remaining five linearly independent combinations of tensor components of (T_{ij}) form a five-dimensional subspace. They can be constructed by requiring both linear independence and orthogonality of all the members of the set of nine combinations. They are, for instance, given by the five combinations.

$$T_{11} + T_{22} - 2T_{33}, \quad T_{11} - T_{22}, \quad T_{23} + T_{32}, \quad T_{31} + T_{13}, T_{12} + T_{21} \tag{4-11}$$

In this manner we have succeeded in subdividing the set of nine components T_{ij}, all nine of which couple to each other under a general coordinate transformation, into a number of subsets or subspaces whose components transform only among themselves under the same transformation. In order to do so, we had to form special linear combinations of the initial tensor components. This suggests that we have carried out a *generalized coordinate transformation* on the initial tensor components.

Let us reformulate our procedure from this point of view. Consider the nine tensor components T_{ij} as the coordinates t_i of a nine-dimensional vector $\mathbf{t}$ in a coordinate system defined by the nine unit vectors $\mathbf{e}_i (i = 1, \ldots, 9)$. The usual assignment of indices is

$$\begin{array}{rccccccccc} (ij) = & 11 & 22 & 33 & 23 & 31 & 12 & 32 & 13 & 21 \\ i = & 1 & 2 & 3 & 4 & 5 & 6 & 7 & 8 & 9 \end{array} \tag{4-12}$$

4.3 Tensor Symmetry and Linear Vector Spaces

An arbitrary rotation (R_{ij}) in real space results in the transformation in nine-space

$$t_i' = \sum_{j=1}^{9} \mathbb{R}_{ij} t_j \qquad (4\text{-}13)$$

where, following Eq. (3-18), the nine-dimensional rotation $\mathbb{R}_{ij}$ is given by

$$\mathbb{R}_{ij} = R_{(i)} R_{(j)} \qquad (4\text{-}14)$$

with the symbols (i) and (j) standing for the two-index assignments corresponding to i and j according to Eq. (4-12). If $(\mathbb{R}_{ij})$ has nonvanishing entries throughout its nine-by-nine array, it will connect all t_j with all t_i'. However, we have already shown by explicit construction that there exists another coordinate system $(\bar{e}_i)$, in which the vector has the components $\bar{t}_i$, such that under an arbitrary transformation

$$\bar{t}_i' = \sum_j \bar{\mathbb{R}}_{ij} \bar{t}_j \qquad (4\text{-}15)$$

the components $\bar{t}_i$ transform among each other in groups of one, three, and five. The basic vectors (e_i) and $(\bar{e}_i)$ are connected by a linear transformation

$$\bar{e}_i = \sum_j \mathbb{D}_{ij} e_j \qquad (4\text{-}16)$$

whose form we know explicitly from the results of Eqs. (4-2), (4-10), and (4-11) to be

$$(\mathbb{D}_{ij}) = \begin{pmatrix} \frac{1}{\sqrt{3}} & \frac{1}{\sqrt{3}} & \frac{1}{\sqrt{3}} & 0 & 0 & 0 & 0 & 0 & 0 \\ \frac{1}{\sqrt{6}} & \frac{1}{\sqrt{6}} & \frac{-2}{\sqrt{6}} & 0 & 0 & 0 & 0 & 0 & 0 \\ \frac{1}{\sqrt{2}} & \frac{-1}{\sqrt{2}} & 0 & 0 & 0 & 0 & 0 & 0 & 0 \\ 0 & 0 & 0 & \frac{1}{\sqrt{2}} & 0 & 0 & \frac{1}{\sqrt{2}} & 0 & 0 \\ 0 & 0 & 0 & 0 & \frac{1}{\sqrt{2}} & 0 & 0 & \frac{1}{\sqrt{2}} & 0 \\ 0 & 0 & 0 & 0 & 0 & \frac{1}{\sqrt{2}} & 0 & 0 & \frac{1}{\sqrt{2}} \\ 0 & 0 & 0 & \frac{1}{\sqrt{2}} & 0 & 0 & \frac{-1}{\sqrt{2}} & 0 & 0 \\ 0 & 0 & 0 & 0 & \frac{1}{\sqrt{2}} & 0 & 0 & \frac{-1}{\sqrt{2}} & 0 \\ 0 & 0 & 0 & 0 & 0 & \frac{1}{\sqrt{2}} & 0 & 0 & \frac{-1}{\sqrt{2}} \end{pmatrix} \quad (4\text{-}17)$$

The square root denominators have been introduced in Eq. (4-17) to make $\mathbb{D}_{ij}$ a *normalized* orthogonal transformation. The arbitrary transformation $(\overline{\mathbb{R}}_{ij})$ is obtained from $(\mathbb{R}_{ij})$, by analogy with the result of Problem 3-8, by carrying out the transformation

$$\overline{\mathbb{R}} = \mathbb{D}\,\mathbb{R}\,\mathbb{D}^{-1} \quad (4\text{-}18)$$

By our construction we have assured that $\overline{\mathbb{R}}$ has *block form*

$$(\overline{\mathbb{R}}_{ij}) = \begin{pmatrix} \boxed{1} & 0 & 0 & 0 & 0 & 0 & 0 & 0 & 0 \\ 0 & & & & & & 0 & 0 & 0 \\ 0 & & & & & & 0 & 0 & 0 \\ 0 & & & 5\times 5 & & & 0 & 0 & 0 \\ 0 & & & & & & 0 & 0 & 0 \\ 0 & & & & & & 0 & 0 & 0 \\ 0 & 0 & 0 & 0 & 0 & 0 & & & \\ 0 & 0 & 0 & 0 & 0 & 0 & & 3\times 3 & \\ 0 & 0 & 0 & 0 & 0 & 0 & & & \end{pmatrix} \quad (4\text{-}19)$$

4.3   Tensor Symmetry and Linear Vector Spaces   37

In this block form, nonzero entries occur only within each block. As a result, the transformation connects only components within each subspace. This holds for any arbitrary transformation in real space (R_{ij}).

In the frame of reference given by the basis vectors $\bar{e}_i$ *all* second rank tensors are fully decomposed into their invariant subspaces, and the transformations ($\mathbb{R}_{ij}$) exhibit fully *reduced* form. What have we accomplished by this reduction, in practical terms?

1. It exhibits explicitly the true tensor invariants.

2. It automatically contains the orthogonality of the linear combinations of tensor components that we introduced arbitrarily in Section 4-2; hence, it facilitates finding the relations, if any, between nonvanishing tensor components.

3. It is an important guide for the explicit construction of tensor components and it tells immediately which tensor components are, and which are not, related to each other under arbitrary coordinate transformations (R_{ij}).

4. It reduces the required number of entries in ($\mathbb{R}_{ij}$) from a total of 81 to at most 35, and because of its block form it simplifies significantly the practical problem of carrying out transformations.

5. It can be generalized to include tensors of any rank.

This generalization is straightforward. For instance, for the third rank tensor we introduce a 27-dimensional coordinate system. We then carry out the reduction, that is, find the particular coordinate system ($\bar{e}_i$) with respect to which arbitrary transformations (R_{ij}) in crystal space are represented in this high-dimensional space by transformations involving irreducible blocks of the form given in Eq. (4-19). We already know the structure of some of these blocks. From Eq. (4-1), there exists one one-dimensional block. Since (T_{ijk}) transforms like products of components of three vectors $\mathbf{V}_1, \mathbf{V}_2, \mathbf{V}_3$, and there are three independent vectors that can be constructed out of these, there must exist three three-dimensional blocks. Furthermore, by viewing some of the tensor components as resulting from the combination of the vectors $\mathbf{V}_3$ and ($\mathbf{V}_1 \times \mathbf{V}_2$), we immediately can use the results already obtained for the second rank tensor (T_{ij}). The one-dimensional block, one three-dimensional block, and a five-dimensional block make up three subspaces of exactly the form of Eq. (4-19). Finally, we can use symmetry arguments to classify the remaining subspaces. The nine-dimensional space resulting from $\mathbf{V}_3$ and ($\mathbf{V}_1 \times \mathbf{V}_2$) is *antisymmetric* in the indices 1 and 2. Hence orthogonality requires that the remaining subspaces be *symmetric* in 1 and 2, and they fall into two groups depending on whether the index 3 is symmetric or antisymmetric with indices 1 and 2. The totally symmetric subspace contains ten possible components (which correspond to the ten different products occurring in the expression $(x + y + z)^3$), and since we can construct a totally symmetric vector from $(\mathbf{V}_1 \cdot \mathbf{V}_2) \mathbf{V}_3 + (\mathbf{V}_2 \cdot \mathbf{V}_3) \mathbf{V}_1 + (\mathbf{V}_3 \cdot \mathbf{V}_1) \mathbf{V}_2$, this space of dimension ten reduces to two subspaces of dimensions three and seven. The subspace in which the index 3 is antisymmetric with the (symmetric) indices 1 and 2 must then contain the remaining three-dimensional space. A five-dimensional space of the same symmetry completes all 27 dimensions.

For later reference, the invariant subspaces for tensors up to fourth rank have been listed explicitly in Appendix 3. For simplification, we have introduced there a two-index notation that makes use of the correspondences between indices and index pairs given by Eq. (4-12). The entries in Appendix 3 are not normalized, but the proper normalization constant can be supplied practically by inspection.

The dimensions of the invariant subspaces constructed so far are all odd. This is a general result. It is connected with the fact that the transformations of these subspaces are actually the transformations for the associated Legendre polynomials $P_l^m(\theta)$ with integral l (or, equivalently, for the spherical harmonics $Y_l^m(\theta, \phi)$). For given l, the set of $2l + 1$ polynomials forms a complete set of orthogonal functions transforming like linear combinations of homogeneous coordinate products. Hence each set P_l^m ($m = -l, -l+1, \ldots, l$) describes an irreducible subspace of the rotation group of dimension $2l + 1$. It can be shown that all values of l generate a complete and unique set of irreducible subspaces. Since there is a one-to-one correspondence between a transformation (R_{ij}) in real space and the transformations $(\overline{\mathbb{R}}_{ij}^l)$ of each irreducible subspace, the transformations $(\overline{\mathbb{R}}_{ij}^l)$ are said to be *irreducible matrix representations* of the rotation group element R. The set of all transformations (R_{ij}) is the three-dimensional irreducible representation of this group.

4.4. INVARIANTS OF TRANSFORMATIONS

The explicit form of a transformation matrix $(\overline{\mathbb{R}}_{ij})$ depends on the frame of reference with respect to which the operation R is specified. However, in common with tensors, transformation matrices also have invariants that are independent of this explicit form and of the frame of reference. The invariant of most interest is the *trace* of the transformation matrix, defined as the sum of the matrix elements along the diagonal. As shown in Eq. (4-2), the same invariant occurs in tensors. Furthermore, the trace characterizes the transformation operation R sufficiently that it has received the name *character*, and is designated by the symbol χ. Since any coordinate transformation not involving an inversion can be represented by a simple rotation about a properly chosen axis, χ is only a function of the angle of rotation $\emptyset$. $\chi(\emptyset)$ is easily evaluated explicitly for the irreducible representations of R discussed in Section 4-3. Under a rotation though an angle $\emptyset$ around the polar axis, the set of spherical harmonics $Y_l^m(\theta, \phi)$ for given l transforms according to the diagonal matrix

4.4 Tensor Symmetry and Linear Vector Spaces

$$(\mathbb{R}_{ij}) = \begin{pmatrix} e^{il\phi} & & & & \\ & e^{i(l-1)\phi} & & & \\ & & \cdot & & \\ & & & \cdot & \\ & & & & e^{-i(l-1)\phi} \\ & & & & & e^{-il\phi} \end{pmatrix} \qquad (4\text{-}20)$$

Hence, we obtain the character $\chi_{(2l+1)}(\phi)$ of the irreducible representation of dimension $(2l+1)$ as

$$\chi_{(2l+1)}(\phi) = \frac{\sin\left(l + \frac{1}{2}\right)\phi}{\sin\frac{1}{2}\phi} \qquad (4\text{-}21)$$

If the operation R includes an inversion, Eq. (4-21) holds as it stands or is multiplied by (-1), depending on whether it stands for a polar (odd) or an axial (even) representation. The results of Eq. (4-21) are simple expressions. Thus, for $l = 0$, $\chi_1(\phi) = 1$; for $l = 1$, $\chi_3(\phi) = 1 + 2\cos\phi$, and $\chi_5(\phi) = -1 + 2\cos\phi + 4\cos^2\phi$. In fact, all characters of Eq. (4-21) can be written as power series in $\cos\phi$. Table A-4-1 of Appendix 4 lists the characters $\chi_{(2l+1)}$ up to $l = 8$ in this form.

The transformation matrices for a given tensor describing arbitrary transformations R also are representations of the group of rotations and rotation-inversions. The discussion in Section 4-3 has shown that these matrices can be broken down into block form; therefore they contain a number of irreducible representations of this group. Since characters are additive, it is easy to see that the character of such a *reducible* representation must be the sum of the characters of the irreducible representations contained in it. Furthermore, this decomposition is unique.

This *principle of composition of characters* offers an elegant method for determining the irreducible representations (blocks) contained in the transformation matrices for a given tensor, without having to construct the specific transformation leading to block form (as was done in the last section for T_{ij}). In case this specific construction is needed, the character analysis is still very useful in guiding the construction of the various subspaces.

In order to apply this principle we must establish the rules for determining the characters of the transformation matrices of tensors. Clearly a scalar always transforms into itself, so that its character is $\chi_{scalar} = 1$. A vector transforms under the scheme of Eq. (3-20), so that $\chi_{vector} = 1 + 2\cos\phi$. A general second rank tensor transforms like the product of two vectors, or

$$\chi_{3\times 3}(\phi) = (1 + 2\cos\phi)^2 \tag{4-22}$$

and, obviously, a general tensor of mth rank has a character

$$\chi_{3^m}(\phi) = (1 + 2\cos\phi)^m \tag{4-23}$$

Table A-4-2 gives the explicit form of a few of such tensor characters. Once this form is known, the decomposition of the tensor character into its irreducible components is straightforward. For instance, it can be carried out by systematically eliminating from the given character the characters of the irreducible representations of highest dimensionality, one after another. Thus, a tensor of third rank has a character

$$\chi_{3\times 3\times 3}(\phi) = 1 + 6\cos\phi + 12\cos^2\phi + 8\cos^3\phi \tag{4-24}$$

The irreducible representation of the highest dimension occurring in it is χ_7, and since, according to Table A-4-1, the coefficient of $\cos^3\phi$ in χ_7 is 8, χ_7 is contained once in Eq. (4-24). Hence we can write

$$\chi_{3\times 3\times 3}(\phi) = 2 + 10\cos\phi + 8\cos^2\phi + \chi_7(\phi)$$

Comparing again with Table A-4-1, we see that χ_5 is contained twice in this expression and we can subtract out $2\chi_5$. In this manner all irreducible representations can be extracted from the tensor character in successive steps. The full decomposition for the third rank tensor is

$$\chi_{3\times 3\times 3}(\phi) = \chi_1(\phi) + 3\chi_3(\phi) + 2\chi_5(\phi) + \chi_7(\phi) \tag{4-25}$$

This agrees completely with the analysis of the subspaces of a third rank tensor carried out in Section 4-3. Similar decompositions of other tensor characters are listed in Table A-4-3.

Tables A-4-2 and A-4-3 also contain the characters and the irreducible decomposition of tensors of various intrinsic symmetries. These characters are derived by making use of the formulas for so-called *symmetric product representations*. For example, let us construct the character of the transformation of a symmetric second rank tensor (T_{ij}). If i and j run from 1 to 3, the tensor has six components transforming like the six products (without numerical coefficients) in the polynomial $(x + y + z)^2$. The basic transformation for each index is the transformation of the coordinates themselves and if this transformation has the diagonal elements a, b, c, then its character is $\chi_3(\phi) = (a + b + c)$. The character of the transformation for the symmetric tensor is given by $\chi_{(3\times 3)_s} = a^2 + b^2 + c^2 + ab + ac + bc$ and this expression can be rearranged to give

$$\frac{1}{2}(a^2 + b^2 + c^2) + \frac{1}{2}(a + b + c)^2$$

The second term is just the square of $\chi_3(\phi)$. The first is the character $\chi_3(2\phi)$ describing two consecutive applications of the same transformation. Hence we obtain

$$\chi_{(3\times 3)_s}(\phi) = \frac{1}{2}\chi_3(2\phi) + \frac{1}{2}\chi_3^2(\phi) \tag{4-26}$$

This result holds equally well for the symmetric product representation derived from an arbitrary representation $\chi_I(\phi)$. For example, if in (T_{ij}) both i and j run from 1 to 6, and $T_{ij} = T_{ji}$, then the tensor transforms with a representation whose character is

$$\chi_{(3\times 3)_s \times (3\times 3)_s}(\phi) = \frac{1}{2}\chi_{(3\times 3)_s}(2\phi) + \frac{1}{2}\chi_{(3\times 3)_s}^2(\phi) \tag{4-27}$$

Other representations appearing in Tables A-4-2 and A-4-3 are obtained simply as products of component representations. For instance, a third rank tensor (T_{ijk}) with the first index pair symmetric has the representation

$$\chi_{(3\times 3)_s \times 3}(\phi) = \chi_{(3\times 3)_s}(\phi) \cdot \chi_3(\phi) \tag{4-28}$$

For completeness, these tables also include the characters of multiple totally symmetric product representations (Table A-4-4), and of the totally antisymmetric product representations (Table A-4-5).

While most of this discussion has been in terms of rotations ϕ, it is important to recall again that for a rotation–inversion $\bar{\phi}$ we may have the two possibilities

$$\chi(\bar{\phi}) = \pm\chi(\phi) \tag{4-29}$$

depending on whether the transformation is polar or axial, for even-rank, or axial or polar for odd-rank tensors.

4.5. CRYSTAL SYMMETRY

The symmetry of crystals differs from the symmetry discussed in the last sections in that the set of symmetry operations is finite and includes only a small number of rotations (or rotation-inversions) through well-defined angles about a few axes. Consequently, the requirements that must be met by any quantity that is to be invariant become much less stringent. We expect that the irreducible subspaces derived for the group of all rotations will break up into smaller subspaces specific to each crystal symmetry group. We also expect that the group of transformations of a tensor corresponding to the symmetry group allows more invariants.

The group character of the rotation group is given by the set $\chi(\phi)$ for all ϕ. When the number of symmetry operations is finite, the group character is the set $\chi(\phi_i)$, where the rotations ϕ_i designate all allowed symmetry operations. For instance, for the group (32), with the six elements $1, 3, 3^2, 2, 2', 2''$, the corresponding angles are $\phi = 0°, 120°, 240°, 180°, 180°, 180°$. The group character for the transformation of a general second rank tensor in this group is the set

$$\chi_{3\times 3} = 9, 0, 0, 1, 1, 1$$

In order to analyze this transformation for its irreducible subspaces, we must know the irreducible representations of the group (32). The results of group theory are that this group has *three* irreducible representations, labeled A_1, A_2, and E, and given by the following set of matrices.

Element	1	3	3^2	2	$2'$	$2''$
A_1:	(1)	(1)	(1)	(1)	(1)	(1)
A_2:	(1)	(1)	(1)	(-1)	(-1)	(-1)
E:	$\begin{pmatrix} 1 & 0 \\ 0 & 1 \end{pmatrix}$	$\begin{pmatrix} \frac{-1}{2} & \frac{\sqrt{3}}{2} \\ \frac{-\sqrt{3}}{2} & \frac{-1}{2} \end{pmatrix}$	$\begin{pmatrix} \frac{-1}{2} & \frac{\sqrt{3}}{2} \\ \frac{\sqrt{3}}{2} & \frac{-1}{2} \end{pmatrix}$	$\begin{pmatrix} 1 & 0 \\ 0 & -1 \end{pmatrix}$	$\begin{pmatrix} \frac{-1}{2} & \frac{\sqrt{3}}{2} \\ \frac{-\sqrt{3}}{2} & \frac{1}{2} \end{pmatrix}$	$\begin{pmatrix} \frac{-1}{2} & \frac{\sqrt{3}}{2} \\ \frac{\sqrt{3}}{2} & \frac{1}{2} \end{pmatrix}$

4.5 Tensor Symmetry and Linear Vector Spaces 43

Therefore the characters of the irreducible representations are

$$\chi_{A_1} = 1, \quad 1, \quad 1, \quad 1, \quad 1, \quad 1$$

$$\chi_{A_2} = 1, \quad 1, \quad 1, \quad -1, \quad -1, \quad -1$$

$$\chi_E = 2, \quad -1, \quad -1, \quad 0, \quad 0, \quad 0$$

The character $\chi_{3 \times 3}$ can be decomposed into these irreducible characters in only one way:

$$\chi_{3 \times 3} = 2\chi_{A_1} + \chi_{A_2} + 3\chi_E$$

and we conclude that with respect to the correct system of axes in nine-dimensional space the transformation matrix for a second rank tensor in the group (32) breaks up into three one-dimensional blocks and three two-dimensional blocks. Furthermore, in the two one-dimensional blocks labelled A_1, all entries have the value +1, so that the component in question transforms into itself under all symmetry operations of the group (32). Hence there are two invariants, and the general second rank tensor (T_{ij}) has two independent components in the group (32).

This is an example of the general method which applies to all tensors and all crystal groups once we know (a) the group character of the transformations of the tensor in question and (b) the list of all irreducible representations of the groups of crystal symmetry. The first requirement has been answered in Section 4-4; the second is met by listing the totality of the irreducible representations and their group character, and is given in Appendix 5. Details of the notation are also explained in this Appendix. However, one general comment is of importance here. As is apparent from the example given above, the characters for different and distinct symmetry elements may be grouped in *classes* that contain all elements of a given type that are equivalent to each other under another symmetry operation of the group. Thus, in the group (32) the two elements 3 and 3^2 form a class, and similarly the three elements 2, 2′, 2″ are in the same class. Clearly, all elements in the same class have the same character. The simplification obtained by introducing classes of elements is included in the tables of Appendix 5.

In addition, group theory provides a further tool for carrying out the analysis of a tensor representation for its irreducible parts. Characters of different irreducible representations are *orthogonal* in the sense

$$\sum_{\text{classes}} N_C \chi_m(C) \chi_n^*(C) = h \delta_{mn} \tag{4-30}$$

This expression applies to a given group with N_C the number of elements in each class C; $\chi_m(C)$, $\chi_n(C)$ class characters of the mth and nth irreducible representations (the star * is included because some characters are complex); and h the number of elements in the group.

An extension of Eq. (4-30) gives the formal rule for determining how often the mth irreducible representation χ_m is contained in a tensor representation χ:

$$n_m = \frac{1}{h}\left[\sum_{\text{classes}} N_C \chi(C) \chi_m{}^*(C)\right] \tag{4-31}$$

In particular, if we are searching for the invariants of a tensor, or equivalently, for the number of independent tensor components, Eq. (4-31) applies if we use for χ_m the totally symmetric irreducible representation of dimension one having entries of +1 for all symmetry elements. Thus, for instance, for the foregoing example

$$n_{A_1} = \frac{1}{6}[1 \cdot 9 \cdot 1 + 2 \cdot 0 \cdot 1 + 3 \cdot 1 \cdot 1] = 2$$

The other irreducible representations obtained in the earlier decomposition also follow immediately. Thus,

$$n_E = \frac{1}{6}[1 \cdot 9 \cdot 2 + 2 \cdot 0 \cdot (-1) + 3 \cdot 1 \cdot 0] = 3$$

Finally, for proper use of these formulas we must clarify the effect of the parity of the tensor. The rule of Eq. (4-29) applies in determining the group character of the tensor transformation. For instance, let us determine the number of invariants of an *axial* second rank tensor in the group (3m). The group has the classes 1, 2(3), and 3(m), and the group character for the second rank tensor is

$$9, \quad 2(0), \quad 3(-1)$$

This results in the decomposition ($n_{A_1} = 1, n_{A_2} = 2, n_E = 3$), or

$$\chi_{A_1} + 2\chi_{A_2} + 3\chi_E$$

so that an axial second rank tensor has only one invariant. The same result is obtained from Eq. (4-31).

$$n_{A_1} = \frac{1}{6}[1 \cdot 9 \cdot 1 + 2 \cdot 0 \cdot 1 + 3 \cdot (-1) \cdot 1] = 1$$

This single invariant can be found either by the direct inspection method of Section 3-7, or by identifying the invariants using the arguments of Section 4-2. It is given by

$$T_{12} - T_{21}$$

The group theoretical method for determining invariants and the number of independent tensor components is explored further in the problems of this chapter.

Problems

4-1. Show, by properly compounding the vectors identified with tensor indices, that if the third rank tensor is symmetric either in a pair of indices, or in all indices, it has no invariant. Show that if it is antisymmetric in a pair of indices, T_{ijk} has at most nine nonvanishing components, and includes one invariant.

4-2. Show, by properly compounding the vectors identified with tensor indices, that if the fourth rank tensor T_{ijkl} is symmetric in either three or all four of its indices, the tensor has one invariant.

4-3. Show that the fifth rank tensor $T_{ijklm} = T_{jiklm}$ has only three invariants, even though we can construct four scalars from the five vectors $\mathbf{V}_1$, $\mathbf{V}_1$, $\mathbf{V}_3$, $\mathbf{V}_4$, $\mathbf{V}_5$ assigned to the tensor indices.

4-4. Suppose the z-axis is an axis of cylindrical symmetry. Show that a general second rank tensor T_{ij} has three invariants which can be derived from the scalars obtained by combining the vectors $\mathbf{V}_1 = (\mathbf{v}_1, w_1)$ and $\mathbf{V}_2 = (\mathbf{v}_2, w_2)$, where v_i is a vector in the xy plane, and w_i is the z component of $\mathbf{V}_i$.

4-5. Apply the method of Problem 4-4 to a general third rank tensor. Show that it has seven invariants, and construct an explicit list of all nonvanishing tensor components.

4-6. A combination of tensor components orthogonal to Eq. (4-1) is given by

$$T_{231} + T_{321} + T_{312} + T_{132} + T_{123} + T_{213}$$

Show that another orthogonal combination of the same six tensor components *symmetric* in the first two indices is

$$2(T_{231} + T_{321}) - (T_{312} + T_{132}) - (T_{123} + T_{213})$$

Find a third orthogonal combination of the same index symmetry, and then construct the two corresponding orthogonal combinations *antisymmetric* in the first two indices.

4-7. Verify that Eq. (4-6) describes an orthogonal set of linear combinations of tensor components.

4-8. Show from Eq. (4-7) that if the fourth rank tensor (T_{ijkl}) has the intrinsic symmetry $T_{ijkl} = T_{jikl} = T_{ijlk}$, it has only two independent tensor components in an isotropic solid, and use Eq. (4-9) to establish the relations between the nonvanishing components.

4-9. Carry out the transformation indicated in Eq. (4-18) for the arbitrary rotation given by Eq. (3-8), and show that it leads to the block form of Eq. (4-19) by evaluating typical elements of $(\mathbb{R}_{ij})$ that are outside the blocks.

4-10. Verify that the first three subspaces of the third rank tensor listed in Table A-3-2 are constructed by identifying the vectors $(\mathbf{V}_1 \times \mathbf{V}_2)$ and $\mathbf{V}_3$ with the two vectors associated with the entries of Table A-3-1.

4-11 Classify the different subspaces listed in Table A-3-3 according to the symmetry of their indices. (Use the full four-index notation wherever necessary.) Show how the number of independent components is reduced when we introduce the intrinsic symmetry $T_{ij} = T_{ji}$. How many components are expected in a fourth rank tensor symmetric in all four indices?

4-12. Verify selected entries in Table A-4-1 by direct evaluation of Eq. (4-21).

4-13. Show, using Eq. (4-21), that $\chi_{2l+1}(0) = 2l + 1$. This is a good check on the coefficients of the entries in Table A-4-1. Is the result $\chi(0) = $ dimension of representation valid for all representations? Why?

4-14. Verify the entry for the character $\chi_3^4(\phi)$ in Table A-4-2, and decompose it into irreducible components following the method outlined in Section 4-4. Check your answer by referring to Table A-4-3.

4-15. Show that the character $\chi_{(3^4)_s}(\phi)$ of Table A-4-2 follows from the rule for a totally symmetric product representation given in Table A-4-4.

4-16. Construct the matrices (R_{ij}) transforming a vector under the group (32). Show that they correspond to the three-dimensional representation composed of the irreducible representations $A_2 + E$ discussed in Section 4-5, and verify this result by character analysis.

4-17. Extend the analysis of a second rank tensor given in Section 4-5 for (32) to the group $(\bar{3}m)$. Consider both polar and axial second rank tensors.

4-18. Verify Eq. (4-30) for the irreducible representations of the group $(\bar{6}2m)$.

4-19. Apply Eq. (4-31) to determine the number of independent components of the tensor (T_{ijkl}), with $T_{ijkl} = T_{kl\,ij}$, in the groups (4), $(\bar{6})$, and $(m3)$.

4-20. Apply Eq. (4-31) to show that the number of independent components of the fully symmetric tensor of sixth rank is 4 for the group (23), and 3 for the group (43). How many components exist in isotropic material?

4-21. Analyze the transformation of a general fourth rank tensor for its irreducible representations in the group $(6mm)$. Consider both polar and axial tensors.

4-22. Show that a second rank tensor does not distinguish between cubic symmetry and isotropy. Does this hold also for a third rank and a fourth rank tensor?

Bibliography

H. Jagodzinski, "Crystallography" in *Encyclopedia of Physics*, S. Flügge, Ed., Springer, Berlin (1955), Vol. VII, part 1.

L. D. Landau and E. M. Lifshitz, *Quantum Mechanics*, Addison-Wesley, Inc., Reading, Mass. (1955), Chapter 12.

S. Bhagavantam, *Crystal Symmetry and Physical Properties*, Academic Press, New York (1966), Chapters 4, 6, 7

M. Hamermesh, *Group Theory and Its Applications to Physical Problems*, Addison-Wesley, Inc., Reading, Mass. (1962), Chapters 3, 4, 5.

CHAPTER 5

Electric Polarization

One of the most common examples of anisotropic physics is the response of electrically polarizable crystals to applied electric fields. Experimentally, dielectric constants are found to be large and their anisotropy can be pronounced. Theoretically, the description of this anistropy is considered basic to all of crystal physics and has been widely discussed in the texts. We will summarize here its essential formulation and then develop some of the specific aspects of dielectric physics that result from the anisotropy of the medium. As will be brought out repeatedly in later chapters in connection with other interactions, many of the formal results and especially the discussion of symmetry can be carried over from this particular application to other situations where the response of the medium has the same formal description. In this sense, this chapter exemplifies the symmetric interaction between two vectors.

5.1. FORMULATION OF THE INTERACTION

Induced electric polarization is the response of a material to an electric field **E**. It is described by an electric moment density **P**. At low fields the response is linear in **E** and takes the form

$$P_i = \epsilon_0 \sum_j \alpha_{ij} E_j \tag{5-1}$$

where ϵ_0 is the rationalized mks scale factor, and (α_{ij}) is the *electric susceptibility tensor*. This material response is incorporated fully in Maxwell's equations by introducing the electric displacement **D**,

$$\mathbf{D} = \epsilon_0 \mathbf{E} + \mathbf{P} \tag{5-2}$$

Since according to Eq. (5-1) the vectors **E** and **P** are usually along different directions, the direction of **D** generally will not coincide with that of either **E** or of **P**. With pronounced anisotropy, these directions can differ widely, as exemplified in Fig. 5-1. **D** can be related directly to **E** by using Eq. (5-1) to eliminate **P** in Eq. (5-2). The relation is linear, and if it is defined by

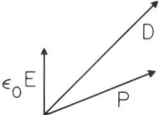

Fig. 5-1 Local relation between **E**, **P**, and **D** in an anisotropic dielectric.

$$D_i = \sum_j \epsilon_{ij} E_j \qquad (5\text{-}3)$$

the *dielectric constant tensor* ϵ_{ij} is given by

$$\epsilon_{ij} = \epsilon_0 (\delta_{ij} + \alpha_{ij}) \qquad (5\text{-}4)$$

The definition of **P** in Eq. (5-1) refers to polarization as an intrinsic volume property. However, observable effects ascribed to **P** are primarily due to $\nabla \cdot \mathbf{P}$ and therefore especially also to surface charge distributions arising from **P**. This implies, for example, that the behavior of finite bodies is largely determined by the influence of the surface and the boundary conditions existing there, and cannot always be inferred from considerations of a volume element in an infinite solid.

For sufficiently strong fields **E**, Eqs. (5-1) and (5-3) may only represent the first term in a series expansion of powers of **E**. Similarly, (α_{ij}) and (ϵ_{ij}) may be functions of other variables of state of the material, such as temperature, magnetic field, and strain. Such complications can be taken into account in a number of ways, some of which will be illustrated in later chapters. In this chapter we will assume (α_{ij}) and (ϵ_{ij}) to be constant tensors fully describing the response of the material to a time-independent or slowly varying electric field **E**.

5.2. SYMMETRY CONSIDERATIONS

Both (α_{ij}) and (ϵ_{ij}) are second rank tensors. Furthermore, they are *symmetric* tensors because the electrostatic field is a conservative field. A sufficient condition for this intrinsic symmetry may be obtained directly from the change in electrostatic free energy density accompanying changes in the fields. From Maxwell's equations, the differential of electrostatic energy density is $dU = \sum_i E_i dD_i$. We can then define an electrostatic free energy $F = U - \sum_i E_i D_i$ whose differential

$$dF = -\sum_i D_i dE_i \tag{5-5}$$

considering F a function of $\mathbf{E}$ is perfect. Consequently, the second cross-derivatives of F are equal, implying that

$$\partial D_i/\partial E_j = \partial D_j/\partial E_i \tag{5-6}$$

and therefore that $\epsilon_{ij} = \epsilon_{ji}$.

This derivation gives only a sufficient condition on (ϵ_{ij}), since it relies on the construction of a free energy density

$$F = -\frac{1}{2} \sum_{i,j} \epsilon_{ij} E_i E_j \tag{5-7}$$

Because of the double summation, all antisymmetric contributions in Eq. (5-7) are automatically suppressed. Any antisymmetric terms in (ϵ_{ij}) might become apparent by considering directly the forces or torques on a volume element having an electric moment P.

Thus, for rotations around the z axis the pertinent component of torque $\mathbf{T}$ is

$$T_3 = (\mathbf{P} \times \mathbf{E})_3 = \epsilon_0 \left[(\alpha_{11} - \alpha_{22}) E_1 E_2 + \alpha_{12} E_2{}^2 - \alpha_{21} E_1{}^2 \right] \tag{5-8}$$

in a system of axes fixed in the crystal. As the volume element is turned around the z axis, this is equivalent to turning the electric field $\mathbf{E} = (E \cos \phi, E \sin \phi, 0)$ in the opposite direction in the coordinate system fixed in the crystal. The net work done by the sample in a full rotation must vanish. This requires that

$$\int_0^{2\pi} T_3 d\phi = 0 = \epsilon_0 E^2 \int_0^{2\pi} \left[(\alpha_{11} - \alpha_{22}) \cos \phi \sin \phi + \alpha_{12} \sin^2 \phi - \alpha_{21} \cos^2 \phi \right] d\phi$$

and therefore again $\alpha_{12} = \alpha_{21}$ and $\epsilon_{12} = \epsilon_{21}$. This traditional derivation is not fully satisfactory either because it cannot be carried over directly to a finite sample where the field **E** will not necessarily remain constant during such rotation. An extension of the proof to physically fully realizable situations is discussed in Section 5-3.

With the given intrinsic symmetry, we can now specify the effect of crystal symmetry on (ϵ_{ij}) and (α_{ij}). Since the crystal symmetry of second rank tensors has already been treated extensively in Chapters 3 and 4 (and in their problems), we need only summarize the results here. The tensor schemes of (ϵ_{ij}) and (α_{ij}) in the various crystal groups are reduced to the following five arrangements.

$$\text{cubic} \begin{pmatrix} \alpha_{11} & 0 & 0 \\ 0 & \alpha_{11} & 0 \\ 0 & 0 & \alpha_{11} \end{pmatrix} \qquad \begin{matrix} \text{hexagonal} \\ \text{tetragonal} \\ \text{rhombohedral} \end{matrix} \begin{pmatrix} \alpha_{11} & 0 & 0 \\ 0 & \alpha_{11} & 0 \\ 0 & 0 & \alpha_{33} \end{pmatrix}$$

$$\text{orthorhombic} \begin{pmatrix} \alpha_{11} & 0 & 0 \\ 0 & \alpha_{22} & 0 \\ 0 & 0 & \alpha_{33} \end{pmatrix} \qquad \text{monoclinic} \begin{pmatrix} \alpha_{11} & \alpha_{12} & 0 \\ \alpha_{12} & \alpha_{22} & 0 \\ 0 & 0 & \alpha_{33} \end{pmatrix}$$

$$\text{triclinic} \begin{pmatrix} \alpha_{11} & \alpha_{12} & \alpha_{13} \\ \alpha_{12} & \alpha_{22} & \alpha_{23} \\ \alpha_{13} & \alpha_{23} & \alpha_{33} \end{pmatrix}$$

Except for the cubic and triclinic groups, these explicit schemes apply in the special coordinate systems singled out by *crystal symmetry*. We have chosen to make the z axis the principal symmetry axis. As expected from the discussion in Chapter 4, higher than twofold axes of rotation lead to isotropy of the second rank tensor in the plane normal to this axis. Furthermore, the dielectric tensor connects two polar vectors and is therefore also a *polar tensor*. It does not distinguish between rotations and rotation-inversions. As a consequence of these two properties, the dielectric tensor cannot distinguish between symmetry groups belonging to the same crystal system.

Because of their intrinsic symmetry, these schemes of constants can be diagonalized in some coordinate system. In their diagonal form, of course, the complete specification of (ϵ_{ij}) or (α_{ij}) must include, in addition to the values of the three diagonal elements, the direction cosines connecting this principal axis system to the conventional crystal symmetry axes.

5.3. DEPOLARIZATION FIELDS AND SHAPE ANISOTROPY

When a sample of finite dimensions is introduced in a region of a uniform electric field $\mathbf{E}^0$, the induced polarization gives rise to surface charges that also contribute to the electric field, both inside the sample and in its surrounding region. In general, the local polarization will be position dependent, and is the result of a self-consistent adjustment of the polarization throughout the medium. The polarization assumes a value everywhere to produce precisely the local electric field necessary to give rise to the existing polarization through Eq. (5-1). This self-consistency, and hence the distribution of polarization, is a sensitive function of the shape and orientation of the sample relative to the external field $\mathbf{E}^0$. For arbitrary sample shape it must be determined anew in each specific case by a solution of Maxwell's equations.

A simple general solution exists, however, for all *ellipsoidal* shapes. Self-consistency becomes easy for ellipsoids because a uniform polarization gives rise to a uniform electric field within the body. This is shown by the following derivation. Consider an ellipsoid of uniform charge density ρ. Its surface is described by a second-order equation. As a result, the potential of the charge distribution is given by a second-order function of coordinates

$$\Phi_0 = \frac{\rho}{2\epsilon_0} \sum_{i,j} L_{ij} x_i x_j \tag{5-9}$$

where the constants L_{ij} are chosen such that Φ_0 satisfies the electrostatic boundary conditions at the surface. By superimposing two slightly displaced ellipsoids of opposite charge density, we can derive from Eq. (5-9) the potential of a dipole distribution $\mathbf{P} = (P_1, P_2, P_3)$,

$$\Phi_1 = (1/\epsilon_0) \sum_{i,j} L_{ij} x_i P_j \tag{5-10}$$

The electric field associated with Φ_1 is

$$E_i^c = (-1/\epsilon_0) \sum_j L_{ij} P_j \tag{5-11}$$

For a given $\mathbf{P}$, $\mathbf{E}^c$ is uniform but not necessarily parallel to $\mathbf{P}$. Since it acts in a direction generally opposite to $\mathbf{P}$ and tends to reduce the total field, it is known as a *depolarization field*. A typical relation between $\mathbf{P}$ and $\mathbf{E}^c$ is shown in Fig. 5-2.

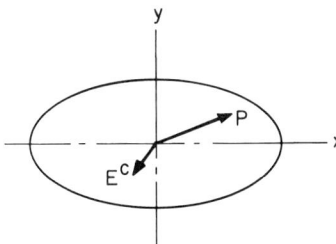

Fig. 5-2 Depolarization field of the shape shown in the outline.

The relation between **P** and $\mathbf{E}^c$ is entirely determined by the set of coefficients L_{ij}. It follows from Eq. (5-9) that the L_{ij} have the intrinsic symmetry $L_{ij} = L_{ji}$, and furthermore, that they transform like products of coordinates under a coordinate transformation. Hence they form a symmetric second rank tensor. Clearly, the tensor is diagonal in the shape principal axis system of the ellipsoid. Its values are known in tabular form for all axis ratios of the ellipsoid. Some simple limiting cases are listed in Table 5-1, and additional examples are treated in the problems. Because the ellipsoid shape is fully determined by two axis ratios, only two of the three values L_i are independent. The third follows from the requirement

$$\sum_i L_i = 1 \qquad (5\text{-}12)$$

Table 5-1
Elements of the Diagonal Shape Depolarization Tensor for Limiting Ellipsoids

	L_1	L_2	L_3
Sphere	$\frac{1}{3}$	$\frac{1}{3}$	$\frac{1}{3}$
Disk in xy plane	0	0	1
Circular cylinder along z axis	$\frac{1}{2}$	$\frac{1}{2}$	0

Once we know the relation of the depolarization field to **P**, as in Eq. (5-11), we can determine the desired self-consistent solution for the state of polarization of the sample in the external field $\mathbf{E}^0$ by constructing the total field

$$\mathbf{E} = \mathbf{E}^0 + \mathbf{E}^c \qquad (5\text{-}13)$$

and using Eqs. (5-1) and (5-11) to solve directly for **P** in terms of $\mathbf{E}^0$. In the *shape principal axis system* (which does not necessarily coincide with that of (α_{ij})) we must solve the equation

$$\sum_j (\delta_{ij} + \alpha_{ij} L_j) P_j = \epsilon_0 \sum_j \alpha_{ij} E_j^0 \qquad (5\text{-}14)$$

Equation (5-14) has the solution, developed in Problem 5-5,

$$P_i = \epsilon_0 \sum_j \alpha_{ij}^e E_j^0 \qquad (5\text{-}15)$$

where the *effective* susceptibilities α_{ij}^e also obey the intrinsic symmetry

$$\alpha_{ij}^e = \alpha_{ji}^e \qquad (5\text{-}16)$$

if, and only if, $\alpha_{ij} = \alpha_{ji}$. Hence we have shown that, at least for all ellipsoids, the net effect of the finite dimensions of the sample on the polarization is merely the replacement of (α_{ij}) by (α_{ij}^e). With such replacement, all arguments using volume elements of an infinite body, such as in Section 5-1, carry over immediately to finite samples.

Equation (5-14) predicts that an ellipsoidal sample composed of isotropic material also shows anisotropic effects. If α is the scalar characterizing the isotropic material, then Eq. (5-14) can be solved directly for **P**:

$$P_i = \epsilon_0 \left(\frac{\alpha}{1 + \alpha L_i} \right) E_i^0 \qquad (5\text{-}17)$$

so that **P** inside the body and the external field $\mathbf{E}^0$ usually have different directions. This shape anisotropy can be avoided only in a spherical sample where all L_i are the same.

5.4. MEASUREMENT OF DIELECTRIC CONSTANTS

The simplest arrangement for measuring the components of the dielectric tensor occurs in the conventional thin parallel plate condenser. Here the electric field direction is dictated by geometry to be normal to the plates, and Table 5-1 shows that the depolarization field is also in this direction. Hence the measurement essentially involves only a single direction. Maintaining the imposed electric field in the presence of the depolarization field, for example, by fixing the potential on the capacitor, requires additional charge on the plates, and thus the dielectric slab increases the capacitance. The ratio of the capacitances of the system with either dielectric or vacuum between the plates is given by the ratio of charge densities at constant voltage

$$D_n/\epsilon_0 E^0 = \epsilon_n/\epsilon_0. \tag{5-18}$$

For condenser plates that are horizontal, Fig. 5-1 describes the relations between the applied field $\mathbf{E}^0$, and $\mathbf{P}$ and $\mathbf{D}$. Obviously D_n, the component of $\mathbf{D}$ along $\mathbf{E}^0$, is larger than $\epsilon_0 \mathbf{E}^0$. ϵ_n is the appropriate diagonal component of (ϵ_{ij}) in a coordinate system one of whose axes lines up with the electric field.

If the direction of the field relative to crystal axes is known, ϵ_n can be expressed in terms of the components of the dielectric tensor in the crystal system by using the transformations discussed in Chapter 3. The whole set of these components is then determined by using a sufficiently large number of dielectric slabs to yield a complete set of independent values of ϵ_n.

If the crystal axes are not known, we may choose an arbitrary system of coordinates to specify the dielectric tensor. In that case the slab directions for independent measurements could be along the x, y, and z axes, and somewhere in each of the xy, the yz, and the xz planes of this system. Once the scheme of coefficients is known, it may then be examined for additional symmetry by transforming it to other frames of reference, as discussed in Chapter 3.

An alternative method of determining (ϵ_{ij}), using only a single sample, relies on the torques and forces experienced by a finite polarizable body in an external field. For a general ellipsoid the theory of these measurements is straightforward. Let us use the shape principal axis system as the frame of reference. Then it follows from Eq. (5-8) that the torque on an electric field $\mathbf{E}^0 = (E^0 \cos \phi, E^0 \sin \phi, 0)$, shown in Fig. 5-3, around the z axis is given by

$$T_3 = -\epsilon_0 E^{0^2} V \left[(\alpha_{11}^e - \alpha_{22}^e) \cos \phi \sin \phi + \alpha_{12}^e (\sin^2 \phi - \cos^2 \phi) \right] \tag{5-19}$$

where the effective susceptibilities of Eq. (5-15) enter to take care of the shape anisotropy. V is the volume of the sample.

The torque T_3 vanishes at the angle ϕ' that also defines the transformation in the xy plane for diagonalizing the two-dimensional submatrix of (α_{ij}^e)

$$\begin{pmatrix} \alpha_{11}^e & \alpha_{12}^e \\ \alpha_{12}^e & \alpha_{22}^e \end{pmatrix} \tag{5-20}$$

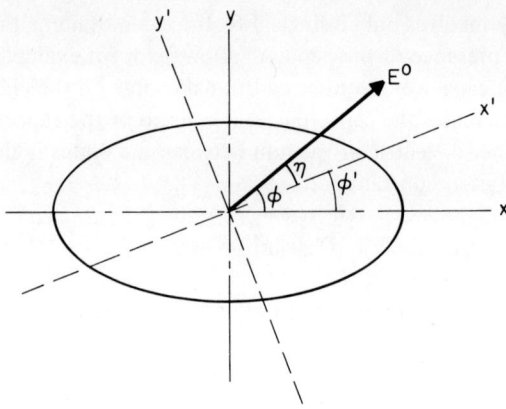

Fig. 5-3 Relation between electric field $\mathbf{E}^0$, the shape principal axis system x, y, and the reference frame x', y' diagonalizing Eq. (5-20).

Hence the equilibrium position of the sample when it is allowed to rotate around the z axis specifies this direction of partial diagonalization of (α_{ij}^e). Furthermore, if we rewrite ϕ as $\phi = \phi' + \eta$, as shown in Fig. 5-3, and keep only first terms in η, then Eq. (5-19) takes the form

$$T_3 = -\epsilon_0 E^{0^2} V(\alpha_{11}^{e'} - \alpha_{22}^{e'}) \eta \qquad (5\text{-}21)$$

where $\alpha_{11}^{e'}$ and $\alpha_{22}^{e'}$ are the principal elements of Eq. (5-20) in the diagonalizing frame of reference. Equation (5-21) predicts that the equilibrium position is stable along the direction of the larger of these two eigenvalues, and that the sample will execute small oscillations about equilibrium with a frequency proportional to $(\alpha_{11}^{e'} - \alpha_{22}^{e'})^{1/2}$.

Similar measurements about the other principal axes of the ellipsoid will give additional information of the same nature about other components of (α_{ij}^e). However, any torque measurements will only yield differences of effective susceptibilities. In order to obtain absolute values it is necessary to complement the torque measurements in uniform field by force measurements in an inhomogeneous field. If the electric field $\mathbf{E}^0$ has a small spatial variation over the region of the sample, the sample is subject to a force $\mathbf{F}$ per unit volume with components

$$F_i = \sum_j P_j \frac{\partial E_i^0}{\partial x_j} = \frac{\partial}{\partial x_i}\left(\frac{\epsilon_0}{2}\sum_{j,k} \alpha_{jk}^e E_j^0 E_k^0\right) \qquad (5\text{-}22)$$

Rather than rely on a good specification of $\mathbf{E}^0$ as a function of position, it is more convenient to compare the force on the sample with the force on an object with known properties having the same dimensions. A metallic ellipsoid serves well for such comparison because its effectively infinite dielectric susceptibilty leads to a simple relation between $\mathbf{P}$ and $\mathbf{E}^0$. From Eq. (5-17), this relation is

$$P_i = (\epsilon_0/L_i)E_i^0 \qquad (5\text{-}23)$$

in the shape principal axis system. If, for example, in this frame of reference the electric field locally has only a large x-component, and this component is only a function of x, then the ratio of forces along x is given by

$$\frac{F_1(\text{diel.})}{F_1(\text{metal})} = \frac{\epsilon_0 \alpha_{11}^e}{\epsilon_0/L_1} = L_1 \alpha_{11}^e \qquad (5\text{-}24)$$

Once the effective susceptibility components are known, the inherent coefficients can be determined by eliminating the effect of shape anisotropy in working backwards through Eqs. (5-15) and (5-14). Obviously, this last step is largely simplified by using a spherical sample. On the other hand, if the material cannot be shaped at will, it may still be well describable by an ellipsoid.

In force and torque measurements on dielectric samples it must be kept in mind that in time charge leakage will tend to depolarize any sample. Since Eqs. (5-19) and (5-22) depend only on the square of the field, such leakage can be reduced by performing the measurements in low-frequency alternating fields. The degree to which this effect can be eliminated depends, of course, on the characteristic times associated with both the mechanical vibration and the leakage conductivity of the material.

5.5. POTENTIAL DISTRIBUTIONS

In a dielectrically anisotropic medium, the static Maxwell's equations, together with Eq. (5-3), lead to Poisson's equation for the electrostatic potential

$$\sum_{i,j} \epsilon_{ij} \frac{\partial^2 \Phi}{\partial x_i \partial x_j} = -\rho(x_i) \qquad (5\text{-}25)$$

where, as usual, $\mathbf{E} = -\nabla\Phi$.

Equation (5-25) can be simplified in two ways. First, the cross-terms are eliminated by performing a transformation to the principal axis system of (ϵ_{ij}). Furthermore, by a subsequent *scale transformation* to new coordinates

$$X_i = x_i/(\epsilon_i)^{1/2} \tag{5-26}$$

the differential equation is formally reduced to that of an isotropic medium, and solutions known in conventional situations can be taken over directly.

However, this is of practical value in only very few cases. Generally, the transformed charge distribution on the right side of Eq. (5-25) (including a possible change of volume because of Eq. (5-26)) does not correspond to standard problems that have already been treated. But even in the absence of free charge, the two consecutive transformations specified above alter the shapes and change the boundary conditions in finite bodies. Finally, another limitation on this approach is imposed by the fact that anisotropy cannot generally be transformed away simultaneously in two or more media.

5.6. COMMENTS ON MAGNETIC SYSTEMS

Although there are formal and real differences between electric and magnetic fields and the corresponding properties of materials, the basic formulation of anisotropic effects is the same for both cases. Many of the detailed results of this chapter can be taken over directly by introducing the appropriate magnetic quantities if we want to deal with para- or diamagnetic effects. For example, since both the magnetization per unit volume **M** and the magnetizing field **H** are axial vectors, the magnetic susceptibility (χ_{ij}) is a polar second rank tensor sharing all formal transformation properties with (α_{ij}). In practice, magnetic susceptibilities are usually orders of magnitude smaller than unity. This simplifies the treatment of finite bodies, because shape anisotropy then becomes an effect of second order. On the other hand, for magnetizable materials near the transition temperature to ferromagnetism the full development of the preceding sections must be taken into account. Actually, because of the absence of the magnetic equivalent of charge leakage, mechanical measurements for determining the properties of magnetic samples have a much more important place in modern experimental work than the corresponding electrical applications discussed in this chapter.

Problems

5-1. Show that in a monoclinic crystal $\alpha_{11}\alpha_{22} > \alpha_{12}^2$, and calculate the maximum angle between **E** and **P** for the two vectors in the xy plane.

5-2. Referring to Fig. 5-2, discuss the amplitude and angle of $\mathbf{E}^c$ as a polarization $\mathbf{P}$ of constant magnitude rotates in the xy plane.

5-3. Show that in elliptic coordinates (e.g., Morse and Feshbach, Section 10.1)

$$x = \frac{a}{2} \cosh \mu \cos \psi \quad y = \frac{a}{2} \sinh \mu \sin \psi$$

the major and minor axes of an ellipse $\mu_0 = $ constant are given by $(a/2) \cosh \mu_0$, $(a/2) \sinh \mu_0$. Solve the electrostatic problem of constant polarization $\mathbf{P} = (P_1, P_2, 0)$ in a cylinder of elliptical cross section and show that the principal components of the depolarization tensor are

$$L_{11} = e^{-\mu_0} \sinh \mu_0, \quad L_{22} = e^{-\mu_0} \cosh \mu_0$$

Verify that Eq. (5-12) is satisfied.

5-4. Verify the result of Table 5-1 for a thin disk by explicit application of electrostatic boundary conditions at the disk surfaces to relate the internal and external fields.

5-5. (a) Show that the effective susceptibility (α_{ij}^e) of Eq. (5-15) can be formally expressed in terms of α and L as

$$\alpha^e = (I + \alpha L)^{-1} \alpha$$

where I is the diagonal unit tensor.
(b) By writing index symmetry symbolically as $A = \widetilde{A}$, and by using the properties $\widetilde{AB} = \widetilde{B}\widetilde{A}$, $\widetilde{A}^{-1} = \widetilde{A^{-1}}$, show that $\alpha^e = \widetilde{\alpha^e}$ if and only if $\alpha = \widetilde{\alpha}$.

5-6. Determine α in terms of L and α^e, and show that $L = (\alpha^e)^{-1} - \alpha^{-1}$.

5-7. Given that in the shape principal axis system

$$L = \begin{pmatrix} 0.3 & 0 & 0 \\ 0 & 0.5 & 0 \\ 0 & 0 & 0.2 \end{pmatrix}$$

and that in the crystal system whose xy axes are rotated $+30°$ with respect to shape axes around the common z axis

$$\alpha = \begin{pmatrix} 2 & 0 & 0 \\ 0 & 4 & 0 \\ 0 & 0 & 4 \end{pmatrix}$$

determine α^e in the system of shape axes.

5-8. Show that the energy increase due to the presence of a dielectric body in a uniform externally fixed electric field $\mathbf{E}^0$ is (see Smythe, Sections 3.12, 3.13)

$$\delta U = - \int (\mathbf{E}^0 \cdot \delta \mathbf{P}) \, dv$$

and calculate that the energy density of a thin anisotropic plate in the xy plane for an external field lying in the xz plane and making an angle θ with the z axis is given by

$$-\frac{1}{2}\epsilon_0 E^{0^2} \frac{[\alpha_{11} + \alpha_{11}\alpha_{22} - \alpha_{12}^2]\sin^2\theta + 2\alpha_{12}\sin\theta\cos\theta + \alpha_{22}\cos^2\theta}{1 + \alpha_{22}}$$

5-9. Specify the minimum number of slabs, and their orientation, needed to determine (ϵ_{ij}) of a monoclinic crystal (a) if the principal direction is known; (b) if it is not known.

5-10. Use the results of Chapter 4 to estimate the dielectric constant of polycrystalline material made up of randomly oriented crystallites with the dielectric susceptibility tensor of Problem 5-7. Considering the correct boundary conditions for **E** and **D** between crystallites, is this result exact?

5-11. Solve Poisson's equation in a rhombohedral medium for a point charge Q at the origin, and for a point dipole **P** at the origin inclined to the principal axis, by use of the methods of Section 5-5.

5-12. Ferroelectric crystals exhibit a spontaneous electric dipole moment.
 (a) Show that ferroelectricity can exist only in crystals not having $\bar{1}$ as a symmetry element.
 (b) Determine the possible directions of the spontaneous polarization in crystals with symmetry ($4mm$), (2), and (m).

Bibliography

J. F. Nye, *Physical Properties of Crystals,* Oxford Univ. Press, London and New York (1957), Chapter 4.

C. Kittel, *Introduction to Solid State Physics,* 4th ed., Wiley, New York (1971), Chapter 13.

W. R. Smythe, *Static and Dynamic Electricity,* McGraw-Hill, New York (1950), Chapter 4.

R. M. Morse and H. Feshbach, *Methods of Theoretical Physics*, McGraw-Hill, New York (1953), Chapters 10, 11.

CHAPTER 6

Magnetic Symmetry

The point symmetry operations of crystals considered so far have been rotations or rotation–inversions. With respect to these symmetry operations we have classified tensor interactions to be either polar or axial, depending on the two possible ways the vectors or combinations of vectors associated with the interaction transform under space inversion. Field vectors, such as the electric field **E** and the magnetic field **H**, also show characteristic behavior under *time inversion*. Since through Maxwell's equation **E** is related to a charge density, while **H** derives from currents, we expect that under time reversal these fields will show the symmetry

$$\mathbf{E}(-t) = \mathbf{E}(t) \qquad (6\text{-}1a)$$

$$\mathbf{H}(-t) = -\mathbf{H}(t) \qquad (6\text{-}1b)$$

By ignoring the effect of time reversal on crystal symmetry, we have, in effect, assumed that all geometrical properties of the crystal involving scalars, such as matter and charge density, or vectors like microscopic electric dipole moments, are invariant under such operation. As far as the crystal is concerned, time inversion is a trivial operation

$$\rho(r, -t) = \rho(r, t) \qquad (6\text{-}2a)$$

$$p_i(r, -t) = p_i(r, t) \qquad (6\text{-}2b)$$

From the point of view of possible interactions, however, such symmetry is not trivial. Its existence permits effects arising from a free energy of the form of Eq. (5-7), since the symmetry $\epsilon_{ij}(-t) = \epsilon_{ij}(t)$ leaves the energy time invariant. On the other hand, it prohibits such effects as

$$F = -\frac{1}{2} \sum_{i,j} \gamma_{ij} E_i H_j \qquad (6\text{-}3)$$

as long as the crystal symmetry requires $\gamma_{ij}(-t) = \gamma_{ij}(t)$, because in that case we cannot form a time-invariant expression for the energy F.

Time reversal may not be a trivial crystal symmetry wherever the distribution of local magnetic moments or spins enters in defining the crystal structure. Magnetic moments and angular momenta are vectors associated with currents and therefore obey the same transformation rule as **H** in Eq. (6-1b). If the microscopic magnetic moment is given by **m**, then as shown in Fig. 6-1,

$$\mathbf{m}(r,-t) = -\mathbf{m}(r,t) \qquad (6\text{-}4)$$

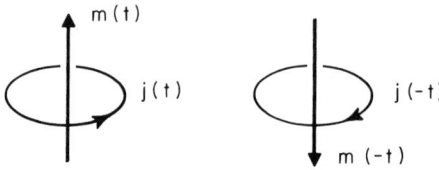

Fig. 6-1 Effect of time reversal $t \to -t$ on a magnetic moment **m** and its generating current j.

This behavior of local vectors associated with the crystal structure has an important bearing on crystal symmetry, and leads to an enlarged set of point symmetry groups. Because the additional symmetry operation arises primarily in connection with magnetic vectors, the new groups are called *magnetic groups*. This chapter extends the methods of Chapters 2, 3, and 4 to magnetic symmetry and discusses its inclusion in the vector–vector interaction of Chapter 5.

6.1. TIME REVERSAL

The foregoing discussion has brought out that time reversal enters into considerations of symmetry in a number of different ways. First, the fields involved in the excitation and in the related response of the solid show characteristic behavior under time inversion that is independent of the nature of the solid. Secondly, a crystal, in particular one containing a distribution of magnetic moments in each unit cell, may undergo a nontrivial transformation under time inversion because the generating currents associated with the magnetic moments all reverse. As a consequence, the coefficients describing a particular interaction must be compatible with the symmetry of the fields as well as that of the crystal under such a reversal. This rule for constructing coefficient schemes applies for the phenomena of *macroscopically reversible thermodynamics*, where we can also derive the interaction from a time-invariant free energy. *Transport phenomena*, however, though based on principles of microscopic reversibility, do not show time reversal symmetry, and the effect of time-inversion symmetry on the microscopic level shows itself in more subtle ways. This is discussed in Chapters 7 and 8.

With respect to the time reversal operation $\mathcal{R}$, all vectors **v** can be divided into two classes,

$$\mathcal{R}\mathbf{v} = \pm\mathbf{v} \tag{6-5}$$

depending on whether the upper or lower sign of Eq. (6-5) holds. If such vectors are part of the crystal structure, we can recognize three classes of crystal symmetry.

(i) The plus sign in Eq. (6-5) always applies. Here $\mathcal{R}$ is a trivial operation, such as in crystals showing no local magnetic moments. It can exist alone as a symmetry operation.

(ii) The minus sign in Eq. (6-5) applies to some vectors. Here $\mathcal{R}$ is a nontrivial operation in crystals containing local magnetic moments. It cannot exist as a symmetry operation by itself because this would reverse the magnetic moments, leaving the rest of the crystal arrangement unchanged. It can exist, however, in combination with other spatial symmetry elements.

(iii) $\mathcal{R}$ is *not* a symmetry operation of the crystal, either alone or in combination with other elements.

The classes (i) and (iii) are fully described by the 32 point groups already developed in Chapter 2, since in (i) $\mathcal{R}$ can be ignored, and in (iii) $\mathcal{R}$ does not apply. In class (ii) the combination of $\mathcal{R}$ with spatial symmetry elements leads to extended magnetic crystal symmetries and to new types of interactions via such crystals. Class (iii) can also describe a magnetic crystal, except that here the magnetic moments go over into each other under the conventional symmetry operations. Classes (ii) and (iii) together, therefore, form groups of magnetic symmetry.

6.2. AN EXAMPLE OF MAGNETIC SYMMETRY

The derivation of the extended symmetry groups obtained by including the time-inversion operation $\mathcal{R}$ is straightforward. However, since in that derivation, given in Section 6-3, we want to use a rather formal approach, it is helpful to precede it by discussion of a simple example that will illustrate the ideas involved.

Let us consider the symmetry of mass points or vectors attached to the corners of a plane hexagon, as shown in Fig. 6-2. The identical mass points, Fig. 6-2a, allow the full symmetry of the hexagonal group ($6/mmm$), with the 24 symmetry elements listed in Appendix 1. If instead of masses we attach vectors to the corners, perpendicular to the plane of the hexagon and alternating in direction out of and into the plane, the rotational symmetry is reduced to a threefold axis and leads to a group of 12 elements. The precise symmetry group depends on whether the vectors are polar (Fig. 6-2b) or axial (Fig. 6-2c).

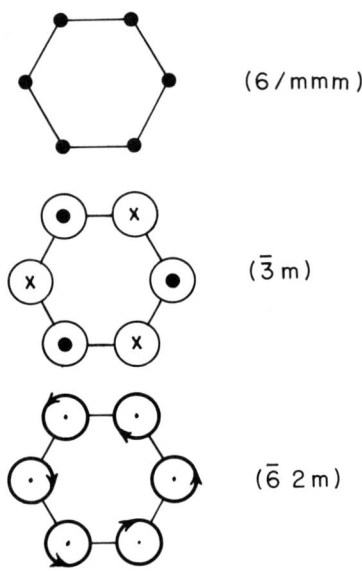

Fig. 6-2 Spatial arrangements and spatial symmetry of mass points, and alternating polar or axial vectors perpendicular to the plane attached to the corners of a planar hexagon.

The reduction in symmetry occurs because the alternating arrangement of the polarity of the vectors makes adjacent corners of the hexagon inequivalent. If these vectors obey the lower relation of Eq. (6-5), so that $\mathcal{R}$ reverses their polarity, the sixfold rotational symmetry can be restored by following an operation such as 6 for the polar vectors by the operation $\mathcal{R}$. The same rule applies to all operations that reverse the vector polarity: by combining them with ℜ the full symmetry is restored. Since the symmetry elements of the full symmetry group ($6/mmm$) either take the vectors into themselves or reverse their polarity, it is evident that the elements that have to be combined with ℜ are the complement of the spatial symmetry group in the full hexagonal group. Thus, the proper magnetic groups for the polar and axial vectors are

$$\text{Polar:} \quad (\bar{3}m) + \Re[(6/mmm) - (\bar{3}m)] \quad (6\text{-}6a)$$

$$\text{Axial:} \quad (\bar{6}2m) + \Re[(6/mmm) - (\bar{6}2m)] \quad (6\text{-}6b)$$

The sets of Eq. (6-6) form groups. They are new symmetry groups of crystals because they contain new symmetry elements. Furthermore, they are realizable because a system and its spin-reversed state can both exist. Finally, these new groups are closely related to the spatial symmetry groups, since whenever ℜ becomes a trivial operation, they must go over into one of the conventional 32 crystallographic groups.

The operator ℜ is commonly designated by the symbol $\underline{1}$. $\underline{1}$ is a symmetry element of order 2 and it commutes with all spatial symmetry operations. Just as for $\underline{1}$ itself, the combined operation of $\underline{1}$ and a spatial element is designated by a bar under the element. Thus: $\underline{1} \cdot 6 = \underline{6}, \bar{3} \cdot \underline{1} = \underline{\bar{3}}$.

6.3. THE MAGNETIC GROUPS

The example of Section 6-2 suggests that the magnetic groups can be generated practically by inspection by attaching the operation ℜ to all spatial symmetry operations that preserve the structure except for the reversal of the polarity of all magnetic vectors. Such a program can be carried out. In this section, however, we will apply a more formal approach that follows the lines of argument used to generate the conventional crystal groups in Chapter 2.

From the two facts that $\Re^2 = 1$ and that ℜ alone cannot be a symmetry operation of the group there follow these easily proven statements for symmetry elements forming a magnetic group:

(i) ℜ occurs only in combination with spatial elements A_C that are of *even* order.

(ii) If the set of symmetry elements contains the pure spatial operation A_S, it cannot also contain $\Re A_S$, and conversely.

(iii) If the set $(A_S, \Re A_C)$ forms a symmetry group composed of two mutually exclusive subsets (A_S) and $(\Re A_C)$, (A_S) alone forms a group.

(iv) The set (A_S, A_C) forms one of the 32 crystal groups.

(v) The number of elements in (A_S) and (A_C) is equal.

These statements suffice to provide a formal procedure for constructing all magnetic groups. Statement (iv) says that we must start with one of the 32 conventional groups. Statement (v) requires that we identify all subgroups of order $n/2$ of the group (of order n). Statement (iii) says that a new group is formed by adding to the subgroup of order $n/2$ the remaining elements of the group, each multiplied by $\mathcal{R}$.

For example, the group $(6/mmm)$ of 24 elements discussed in Section 6-2 has five subgroups of order 12. Hence, the magnetic groups deriving from it are, in addition to Eqs. (6-6a) and (6-6b),

$$(6mm) + \mathcal{R}[(6/mmm) - (6mm)] \tag{6-7a}$$

$$(62) + \mathcal{R}[(6/mmm) - (62)] \tag{6-7b}$$

$$(6/m) + \mathcal{R}[(6/mmm) - (6/m)] \tag{6-7c}$$

The full set of magnetic groups derived in this manner, together with their standard designation, is listed in Appendix 6. Including the conventional groups, the extended symmetry allows 90 distinct magnetic groups.

6.4. MATHEMATICAL DESCRIPTION OF MAGNETIC PROPERTIES

The inclusion of the operator $\mathcal{R}$ in combination with some symmetry elements does not pose any difficulties in the mathematical formulation of crystal properties given in Chapters 3 and 4.

If a tensor changes sign under the application of $\mathcal{R}$, then this change of sign must be included explicitly after a transformation of the tensor component is carried out in terms of the transformation of the corresponding product of coordinates. The procedure is exactly the same as that which has to be followed with axial tensors when the coordinate transformation involves an inversion. One can divide all tensors into two groups with respect to their transformation properties under time inversion, depending on whether the number of tensor indices responsive to time reversal is even or odd.

A simple example will illustrate the procedure. In a *pyromagnetic* crystal a magnetic moment develops due to a change in temperature ΔT

$$M_i = c_i \Delta T \tag{6-8}$$

Let us determine the allowed pyromagnetic coefficients c_i in a crystal of symmetry $(\underline{m})$. This group contains, besides the identity, only the element $\underline{m}$, composed of the product m and $\mathcal{R}$. Since **M** is an axial vector, c_i is an axial tensor. Hence under the operation m given by

6.4 Magnetic Symmetry

$$m = \begin{pmatrix} 1 & 0 & 0 \\ 0 & 1 & 0 \\ 0 & 0 & -1 \end{pmatrix}$$

we obtain

$$c_1' \sim (-1)x' = -x \sim -c_1$$
$$c_2' \sim (-1)y' = -y \sim -c_2 \qquad (6\text{-}9)$$
$$c_3' \sim (-1)z' = z \sim c_3$$

where the extra minus sign denotes the axial nature of c_i. Since **M** also reverses under the operation $\mathcal{R}$, the operation $\underline{m}$ has the effect:

$$c_1' \sim (-1)(-1)\, x' = x \sim c_1$$
$$c_2' \sim (-1)(-1)\, y' = y \sim c_2 \qquad (6\text{-}10)$$
$$c_3' \sim (-1)(-1)\, z' = -z \sim -c_3$$

Hence the equality of the coefficients under this transformation requires that $c_3 = 0$, and that c_1 and c_2 be arbitrary. Thus, in a crystal of symmetry $(\underline{m})$ the magnetization **M** can lie anywhere in the xy plane. On the other hand, according to Eq. (6-9), in a crystal of symmetry (m), **M** must lie along the z axis. The two situations are illustrated in Fig. 6-3.

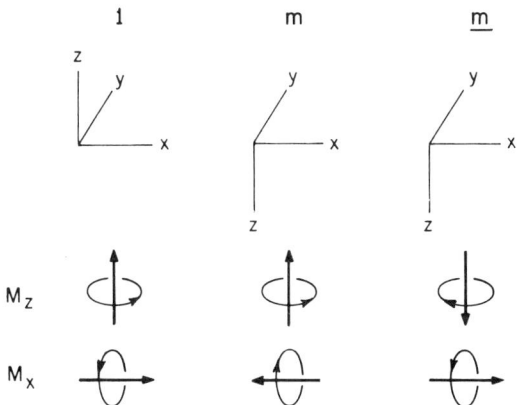

Fig. 6-3 Transformation of the coordinate system and of the components M_z and M_x of the magnetization under the operations m and $\underline{m}$.

The concepts of Chapter 4 can be extended to include magnetic symmetry if the sensitivity to change of sign under $\mathcal{R}$ of the vectors used to construct invariants and other basis vectors of tensors in question is taken into account. Similarly, this change of sign must be included in computing the characters associated with magnetic tensors where the operation in question includes $\mathcal{R}$.

6.5. SOME APPLICATIONS

The extended symmetries are of interest whenever magnetic structures are involved and when the interaction being studied involves magnetic vectors, such as **H** or **M**. Among the three classes of crystals enumerated in Section 6-1 diamagnetic and paramagnetic materials belong to (i). Therefore they cannot show any effects involving odd powers of magnetic vectors (which reverse sign under $\mathcal{R}$). With respect to properties involving even powers of such vectors, the analysis is identical to that involving electric vectors, such as was presented in Chapter 5.

In the classes (ii) and (iii) a magnetic vector may appear an odd number of times. This leads to such properties as pyromagnetism, already discussed in Section 6-4, and the magnetoelectric effect, described by an interaction

$$\Delta M_i = \sum_j \lambda_{ij} E_j \tag{6-11}$$

where $\Delta \mathbf{M}$ is a magnetization density and **E** the electric field. In each case, a magnetic vector appears to first power.

The second rank tensor λ_{ij} differs from the polarizability tensor discussed in Chapter 5 in three respects: (1) It is a general second rank tensor; (2) it is an axial tensor; (3) it is a magnetic tensor reversing with $\mathcal{R}$.

The form of (λ_{ij}) permitted by symmetry in each of the 90 magnetic groups is easily determined by using the methods of Chapter 3. Some examples are given in the problems of this chapter, and complete lists of constants in all symmetries appear in the references. The magnetoelectric polarizability can exist among 58 groups in 11 different schemes, and both longitudinal and transverse couplings are possible. The effect has been observed in crystals, such as Cr_2O_3, that show an antiferromagnetic structure.

In the later chapters the effects of magnetic symmetry on other interactions will be included in the discussion. In all of these applications, however, it must be kept in mind that the magnetic symmetry is considered to be part of the structure. This approach is valid when the externally applied forces and fields do *not* change the symmetry of the isolated crystal, to first order. It is clearly violated when the forces themselves distort the crystal appreciably. In the case of arrangements of magnetic dipoles it requires that these arrangements be dictated predominantly by internal fields, and not destroyed by external influences.

Magnetic Symmetry 69

If this condition is violated, the magnetic ordering must be described as a *property* of the crystal, rather than as part of the symmetry-determining structure. This is the point of view which is normally taken with respect to magnetic and electric polarizability, and in particular with respect to magnetic anisotropy energy. Here we ask for the form of this energy compatible with crystal symmetry, excluding magnetic structure, since this structure can be strongly influenced by externally applied fields.

Problems

6-1. Show that among the 32 crystal groups of Appendix 1 only the groups (1), $(\bar{1})$, (2), (m), $(2/m)$, (4), $(4/m)$, $(\bar{4})$, (3), $(\bar{3})$, (6), $(\bar{6})$, and $(6/m)$ can show ferromagnetism. What is the underlying principle in making the selection? Compare it with that used in Problem 5-12.

6-2. Verify that the set of Eq. (6-6a) forms a group, and that this group has the same multiplication table as the full spatial symmetry group from which it is derived.

6-3. Apply the method of Section 6-2 for generating magnetic groups to the example discussed there if the vectors lie *in* the plane of the hexagon and preserve sixfold rotational symmetry. Show that this leads to three new groups.

6-4. Give the proof for the statements (i)–(v) of Section 6-3.

6-5. Verify the magnetic groups derived from $(6/m)$ in Appendix 6.

6-6. Nickel is a face-centered cubic metal. If it is magnetized along the (111) direction, and if the magnetization is assumed to follow the symmetry of the atomic arrangement, what is its magnetic group?

6-7. Investigate the existence and possible degrees of freedom of pyromagnetism in the symmetries $(\underline{2}/m)$ and $(\underline{2}/\underline{m})$.

6-8. Determine the group character for a magnetic second rank tensor (i.e., one whose components change sign with $\underline{R}$) in the groups (32) and ($\bar{3}m$), and find the number of independent components such a tensor has in these systems. Compare the answer with that of Problem 4-17.

6-9. Extend the analysis of Problem 6-8 to the magnetic groups derived from (32) and ($\bar{3}m$).

6.10. Extend Problem 6-1 to determine the possibility of ferromagnetism in the symmetry groups of Appendix 6.

6-11. Show that magnetoelectricity cannot exist in any symmetry containing $\bar{1}$ as a symmetry element. Use Appendix 6 to determine that this rule applies to 21 groups.

6-12. Determine the scheme of λ_{ij} in the magnetic groups associated with the point group $(6/m)$.

6-13. Show that there is no magnetoelectric coupling in the group ($\underline{4}\,3$).

6-14. (a) Show that in the system (mm) the magnetoelectric tensor of Eq. (6-11) has the form

$$\begin{pmatrix} \lambda_{11} & 0 & 0 \\ 0 & \lambda_{22} & 0 \\ 0 & 0 & \lambda_{33} \end{pmatrix}$$

(b) Show that the generalization of Eq. (5-7) to a magnetoelectric crystal of symmetry (mm) leads to a free energy

$$F = -\frac{1}{2} \sum_i \epsilon_{ii} E_i^2 - \frac{1}{2} \sum_i \mu_{ii} H_i^2 - \sum_i \lambda_{ii} H_i E_i$$

(c) Show that the requirement that F be a negative definite expression imposes the restrictions

$$\lambda_{ii} \leqslant (\epsilon_{ii} \mu_{ii})^{1/2}$$

(d) Show that a more stringent restriction for *paramagnetic* materials is given by

$$\lambda_{ii} \leqslant (\epsilon_0 \mu_0)^{1/2} (\alpha_{ii} \chi_{ii})^{1/2}$$

where α and χ are the electric and magnetic susceptibilities.

(e) In a similar way, find the upper limits on the magnetoelectric constants in the system (mm).

6-15. Following the formulation in Problem 6-14, write down the free energy for the most general magnetoelectric interaction, and use this expression to formulate the converse effect of a polarization produced by an applied magnetic field.

Bibliography

A. V. Shubnikov and N. V. Belov, *Coloured Symmetry*, Pergamon Press, Oxford (1964).

R. R. Birss, *Symmetry and Magnetism*, Wiley, New York (1964).

S. Bhagavantam, *Crystal Symmetry and Physical Properties*, Academic Press, New York (1966), Chapters 5, 7, 15.

M. Hamermesh, *Group Theory and Its Applications to Physical Problems*, Addison-Wesley, Inc., Reading, Mass. (1962), Chapter 2.

L. D. Landau and E. M. Lifshitz, *Electrodynamics of Continuous Media*, Addison-Wesley, Inc., Reading, Mass. (1960), Chapters 3, 4.

CHAPTER 7

Electrical Conduction

Electrical conduction describes the flow of electric charge through crystals in response to applied electric fields. In the regime of fields where the response is linear, the electrical conductivity (σ_{ij}) establishes a linear relation between the current vector and the field vector. It is therefore a second rank tensor similar to those already discussed in Chapters 5 and 6, and many of the results obtained there can be transferred directly to this new application.

However, electrical conduction deserves a separate discussion. First, it is the basic example of an *irreversible* process. The criteria for establishing intrinsic symmetry involve new considerations differing from those used in Chapters 5 and 6. Second, electrical transport phenomena are concerned with problems subject to the special boundary conditions governed by the law of conservation of charge. The role of anisotropy in these specific applications is of practical importance. Finally, electrical transport is also sensitive to the simultaneous application of magnetic fields, and therefore represents a first example of matter response in the presence of more than one applied field.

With minor adjustments, the discussion of electrical properties of this chapter can be taken over fully to describe separately the analogous phenomena of thermal conduction (heat flow due to temperature gradients) and diffusion (mass flow due to concentration gradients). The simultaneous presence of more than one of these processes is treated in Chapter 8.

7.1. THE SYMMETRY OF IRREVERSIBLE PROCESSES

Irreversible processes occur when a system attempts to reach thermodynamic equilibrium from an initial nonequilibrium state characterized by internal variations in temperature, electric potential, or chemical concentrations. Locally, these processes are described by currents **J**, such as thermal, electrical, or mass transport, responding to driving forces **X** (i.e., temperature gradients, potential gradients, or concentration gradients). If a driving force **X** is imposed and maintained externally, there results a steady-state relation between **J** and **X** that is expected to be linear as long as the deviations from equilibrium are small. Hence we can write

$$J_i = \sum_j L_{ij} X_j \qquad (7\text{-}1)$$

where (L_{ij}) is a generalized *conductivity*. Because of Eq. (7-1) (L_{ij}) is a second rank tensor and therefore subject to all the symmetry restrictions already discussed for such tensors.

Independently of crystal symmetry, however, its components are also interconnected by the *Onsager relations*. These relations arise from an inherent symmetry of irreversible processes and express a coupling or correlation between simultaneously occurring transport phenomena. The origin of these relations can be clarified in a number of ways.

Macroscopically, an irreversible process increases the entropy density S of a system. The rate of entropy production dS/dt associated with the flux–force pair **J**, **X** is given by the scalar product of **J** and **X**

$$\frac{dS}{dt} = \sum_i X_i J_i \qquad (7\text{-}2)$$

By eliminating **J** through Eq. (7-1), the entropy production becomes a quadratic function of the X_i. Since it must always be positive, we require the inequality

$$\sum_{i,j} L_{ij} X_i X_j > 0 \qquad (7\text{-}3)$$

The positive-definite character of this expression implies that observable phenomena may depend only on the *symmetric* part of (L_{ij}). Equation (7-3) thus plays a role in irreversible phenomena similar to that of an energy, such as given in Eq. (5-7), for reversible phenomena.

A further argument for the symmetry of L may be obtained by considering the spontaneous fluctuations of the system about equilibrium, and by identifying the law of decay of such fluctuations with the behavior to be expected from an imposed macroscopic deviation from equilibrium. If the system is characterized by a set of parameters A_i, then the entropy change associated with macroscopic changes in the A_i is

$$\frac{dS}{dt} = \sum_i \left(\frac{\partial S}{\partial A_i}\right)_0 \frac{dA_i}{dt} \qquad (7\text{-}4)$$

7.1 Electrical Conduction

The subscript zero refers to the initial equilibrium state. By comparing Eqs. (7-2) and (7-4) we can identify the forces and fluxes associated with the A_i, and the linear relation of Eq. (7-1) now takes the form

$$\frac{dA_i}{dt} = \sum_k L_{ik}\left(\frac{\partial S}{\partial A_k}\right)_0 \tag{7-5}$$

The right side of Eq. (7-5) refers to equilibrium states of the system. Hence we can evaluate ensemble averages of various products, and, for example, prove the result

$$\left\langle A_j\left(\frac{\partial S}{\partial A_k}\right)_0 \right\rangle_{\widehat{ens}} = -k_0 \delta_{jk} \tag{7-6}$$

where k_0 is Boltzmann's constant.

Forming a similar correlation with the left side of Eq. (7-5), and averaging over the long time equivalent to an ensemble average, we find

$$\left\langle A_j \frac{dA_i}{dt} \right\rangle_{\widehat{time}} = -k_0 L_{ij} \tag{7-7}$$

However, if the A's can also be identified with spontaneous fluctuations, it follows from the principle of *microscopic reversibility* that

$$\left\langle A_j \frac{dA_i}{dt} \right\rangle_{\widehat{time}} = \left\langle A_i \frac{dA_j}{dt} \right\rangle_{\widehat{time}} \tag{7-8}$$

Combining Eq. (7-8) with Eq. (7-7), we obtain the central result of the Onsager relations

$$L_{ij} = L_{ji} \tag{7-9}$$

Thus, by combining the requirement of microscopic reversibility with the identification of the law of decay of natural fluctuations with that of irreversible processes, Eq. (7-9) becomes the necessary condition of symmetry coupling the different components of the generalized conductivity. The correct proof involves averages whose detailed specification is much more complex than we have indicated, but our outline suffices to show the logic of the main arguments. It is important to recognize that because of the nature of the proof, Eq. (7-9) applies only to those pairs of forces and currents that satisfy Eq. (7-2).

In the presence of a magnetic field **H**, or of macroscopic velocity-dependent terms which may enter in L, the Onsager relations of Eq. (7-9) must be generalized to include the effect of the time reversal implied in Eq. (7-8) on **H** or **v**:

$$L_{ij}(\mathbf{H}, \mathbf{v}) = L_{ji}(-\mathbf{H}, -\mathbf{v}) \tag{7-10}$$

Based on the most general principles, and independent of special models or mechanisms of transport, Eq. (7-10) occupies a central position in the treatment of all irreversible phenomena.

7.2. ANISOTROPIC CONDUCTION

Electrical transport is described by the *electrical conductivity* tensor (σ_{ij}) in the relation

$$J_i = \sum_j \sigma_{ij} E_j \tag{7-11}$$

or its inverse, the resistivity tensor (ρ_{ij}), in

$$E_i = \sum_j \rho_{ij} J_j \tag{7-12}$$

Both σ and ρ are second rank tensors, and it follows from Eq. (7-9) that in the absence of magnetic fields they are symmetric in their two indices. Hence, the formal anisotropy of σ and ρ is exactly the same as that of ϵ or α. The tensor schemes of σ and ρ in the various crystal groups are therefore given by those already derived in Chapters 3 and 5 for all symmetric second rank tensors.

The major consequence of this anisotropy is that the current **J** does not generally follow the lines of the electric field **E**. **E** and **J** make an angle with each other that is a function of their orientations relative to crystal axes. Their directions coincide only in the planes of isotropy and along the principal axis directions of the crystal.

The choice between Eq. (7-11) and Eq. (7-12) in treating conduction problems is often dictated by the boundary conditions to be satisfied in the specific case under consideration. For example, in the long and thin rod sample commonly used for measuring resistivity, the current direction is fully prescribed by geometry to be along the axis of the rod. If this axis is parallel to x, then the current is $\mathbf{J} = (J, 0, 0)$, and the electric field, given by Eq. (7-12), has the components $\mathbf{E} = (\rho_{11}J, \rho_{21}J, \rho_{31}J)$. Hence **E** is constant but not parallel to **J**. The electrostatic potential Φ associated with **E** derives from the equation

$$\nabla \Phi = -\mathbf{E} \tag{7-13}$$

and, for this particular electric field, is given by

$$\Phi(x,y,z) = -J(\rho_{11}x + \rho_{21}y + \rho_{31}z) \tag{7-14}$$

everywhere inside the conducting rod. Equation (7-14) predicts that the current flow is accompanied by longitudinal as well as transverse potential drops. Figure 7-1 shows a set of equipotentials corresponding to Eq. (7-14). Evidently, voltage

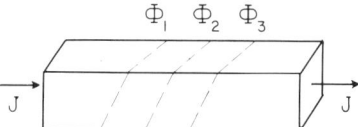

Fig. 7-1 Three equipotential planes in a long thin rod with anisotropic resistivity carrying a constant current density **J**.

probes must be placed very carefully in order to measure only a single one of the three contributions to the potential drop. For measuring the longitudinal voltage in Fig. 7-1, proportional to ρ_{11}, the requirement is that point electrodes be on the same current flow line. As follows from Eq. (7-13) the electric field is normal to the equipotential planes. Its transverse components are maintained during the longitudinal flow of current by an accumulation of equal and opposite charge densities on the pairs of lateral faces of the rod.

In Fig. 7-1 the current flow lines are parallel to the rod axis, and the current density is uniform. This distribution is generally altered at the junction where the current enters or leaves the sample. For example, if the end electrode is made of highly conducting material, it defines an equipotential surface. Hence the inclined equipotentials far from the electrode must shift direction gradually to conform to this boundary condition. Figure 7-2 illustrates such a progression

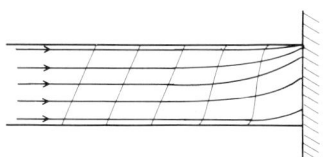

Fig. 7-2 Equipotentials and current flow lines in an anisotropic rod terminated by a perfectly conducting electrode.

of shifting equipotentials as the electrode is approached. It also indicates that the equipotentials must still retain the original inclination near the lateral boundaries in order that the lateral boundary condition of current flow parallel to the faces be satisfied. Figure 7-2 depicts a number of current flow lines in the neighborhood of the electrode. The lines crowd toward one side of the sample, and all current lines enter the end electrode at the same angle.

All of these conclusions follow directly from a qualitative consideration of various boundary conditions. Their quantitative expression requires the solution for the potential $\Phi(x,y,z)$ subject to these conditions. The differential equation for Φ follows by first combining Eqs. (7-11) and (7-13) for an expression of the current

$$J_i = -\sum_j \sigma_{ij} \frac{\partial \Phi}{\partial x_j} \tag{7-15}$$

and then invoking the law of continuity of current

$$\sum_i \frac{\partial J_i}{\partial x_i} = 0 \tag{7-16}$$

The combination of Eqs. (7-15) and (7-16) leads to the equation

$$\sum_{i,j} \sigma_{ij} \frac{\partial^2 \Phi}{\partial x_i \, \partial x_j} = 0 \tag{7-17}$$

which, together with the usual boundary conditions, involving either given equipotentials or given directions of current flow, defines the potential distribution in anisotropic conductors.

7.3. POTENTIAL DISTRIBUTIONS IN ANISOTROPIC CONDUCTORS

The solutions of Eq. (7-17) either can be constructed directly starting from the given differential equation, or they can be related to the solutions of equivalent problems in isotropic media where a great number of techniques for handling specific sets of boundary conditions are already available.

The transformation of Eq. (7-17) to the usual Laplace equation in an isotropic medium requires two steps. First, the off-diagonal terms of (σ_{ij}) are eliminated by rotating the reference frame so that it agrees with the principal axes of (σ_{ij}). Second, by performing a *scale transformation*

$$x_i' = S_i x_i \tag{7-18}$$

the three axis directions are made formally equivalent. The three scale factors S_i contain some arbitrariness, which is removed by requiring that the transformation conserve volume. The equality of volume elements $dx_1' \, dx_2' \, dx_3'$ and $dx_1 \, dx_2 \, dx_3$ leads to the relation

$$S_1 S_2 S_3 = 1 \tag{7-19}$$

The currents and fields transform under Eq. (7-18) as follows

$$J_i' = S_i J_i \tag{7-20a}$$

$$E_i' = (1/S_i) E_i \tag{7-20b}$$

since **J** is proportional to a velocity, and **E** derives from the gradient of an invariant scalar, Eq. (7-13).

Let us determine the scale factors, assuming that the unprimed quantities refer to the principal axis system of (σ_{ij}) such that $J_i = \sigma_{ii} E_i$. Using Eqs. (7-20), the effective conductivity σ_m of the scaled medium is related to the principal conductivities by

$$\sigma_m = J_i'/E_i' = S_i^2 \, (J_i/E_i) = S_i^2 \, \sigma_{ii} \tag{7-21}$$

Inserting these values of S_i in Eq. (7-19) gives the explicit solution for σ_m

$$\sigma_m = (\sigma_{11}\sigma_{22}\sigma_{33})^{1/3} \tag{7-22}$$

and Eq. (7-21) provides the values of the S_i

$$S_1 = \left(\frac{\sigma_{22}\sigma_{11}}{\sigma_{11}^2}\right)^{1/6}, \quad S_2 = \left(\frac{\sigma_{33}\sigma_{11}}{\sigma_{22}^2}\right)^{1/6}, \quad S_3 = \left(\frac{\sigma_{11}\sigma_{22}}{\sigma_{33}^2}\right)^{1/6} \tag{7-23}$$

Hence the transformation Eq. (7-18) is fully determined. As expected, it conserves not only volume, but also electrical current (Problem 7-6). By construction of Eq. (7-21), **E**' and **J**' are parallel, and the boundary conditions on **J** at all surfaces remain unchanged under the transformation.

As a simple example of this transformation technique, let us discuss the field and current distribution of a point source at the origin in an infinite anisotropic medium. In the corresponding scaled isotropic medium the appropriate solution of $\nabla^2 \Phi = 0$ is

$$\Phi(x', y', z') = -\frac{A}{(\sum_i x_i'^2)^{1/2}} \tag{7-24}$$

Hence the electric fields and currents in this medium are

$$E_i' = \frac{Ax_i'}{(\sum_i x_i'^2)^{3/2}}, \quad J_i' = \sigma_m \frac{Ax_i'}{(\sum_i x_i'^2)^{3/2}} \qquad (7\text{-}25)$$

giving a strictly radial electric field and current distribution. The distributions in real space are obtained by inverting the transformations of Eq. (7-18) and (7-20). The current components are

$$J_i = \frac{1}{S_i} \sigma_m \frac{AS_i x_i}{(\sum_i S_i^2 x_i^2)^{3/2}} = \frac{A}{(\sigma_m)^{1/2}} \frac{x_i}{[\sum_i (x_i^2/\sigma_{ii})]^{3/2}} \qquad (7\text{-}26)$$

and since Eq. (7-26) implies the relation

$$J_i/J_j = x_i/x_j \qquad (7\text{-}27)$$

the current flows in radial straight lines away from the origin. On the other hand, the electric field components are

$$E_i = \frac{AS_i^2 x_i}{(\sum_i S_i^2 x_i^2)^{3/2}} = \frac{A}{(\sigma_m)^{1/2}} \frac{1}{\sigma_{ii}} \frac{x_i}{[\sum_i (x_i^2/\sigma_{ii})]^{3/2}} \qquad (7\text{-}28)$$

in agreement, of course, with the relation $J_i = \sigma_{ii} E_i$. It implies that the electric field lines are not radial, but form curves emerging from the origin. This is confirmed by the observation that the equipotentials, given by constant values of Φ in

$$\Phi(x,y,z) = -\frac{A}{(\sigma_m)^{1/2}} \frac{1}{[\sum_i (x_i^2/\sigma_{ii})]^{1/2}} \qquad (7\text{-}29)$$

are surfaces of ellipsoids. The relation between equipotentials, field lines, and currents of this example is shown in Fig. 7-3.

This example only involves a very simple geometrical boundary condition at the origin. In most problems of interest, such as that shown in Fig. 7-2, the scale transformation Eq. (7-18) distorts the geometry of the sample, and the desired solution in the scaled medium may not be among those already known. In constructing new solutions in the scaled isotropic medium, it is helpful to

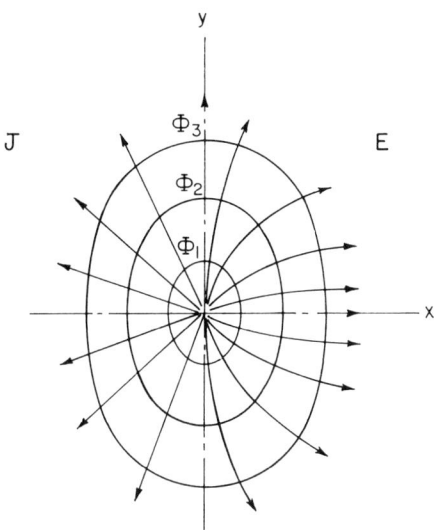

Fig. 7-3 Equipotentials, field lines, and current flow lines for a point source in a medium with $\sigma_{22}/\sigma_{11} = 2$.

remember that under Eq. (7-18) planes remain planar, and that the boundary conditions relating to current flow parallel to and across surfaces are retained.

7.4. FOUR-PROBE MEASUREMENT OF CONDUCTIVITY

Another example of potential distributions in anisotropic media involving simple boundary conditions occurs when the conductivity of the medium is to be determined by four point probes situated on the same surface of the sample.

One of the simplest arrangements for such measurement is shown in Fig. 7-4. A sample of thickness t and large lateral dimensions has on its upper surface four probes equally spaced along a straight line, the two outer ones serving as

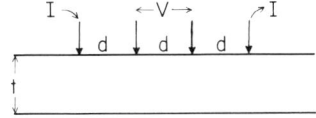

Fig. 7-4 Four-probe arrangement for the measurement of conductivity.

current leads, and the two inner ones for recording the potential drop V. If the material has an isotropic conductivity σ, the current voltage relation takes very simple forms in two limiting cases

$$t \gg d, \quad I = 2\pi d\sigma V \qquad (7\text{-}30a)$$

$$t \ll d, \quad I = (\pi t\sigma/\ln 2)V \qquad (7\text{-}30b)$$

How is the same measurement to be interpreted if the medium is anisotropic? Clearly, the dimensions d and t, and the conductivity σ of Eq. (7-30) are those proper to the equivalent scaled medium. In terms of the transformation of Section 7-3 a measurement of I and V determines the products $d'\sigma_m$ and $t'\sigma_m$. The specific relation of these numbers to the anisotropic conductivity tensor depends on the orientation of the line of probes relative to the principal axes of the medium.

If the line of probes follows a principal direction, for example x_1, then $d'\sigma_m$ can be transformed back into real space by using Eqs. (7-18) and (7-22) to give

$$d'\sigma_m = d(\sigma_{22}\sigma_{33})^{1/2} \qquad (7\text{-}31a)$$

where d is the actual spacing between probes. Similarly, a measurement along a second principal direction in the surface yields the conductivity $(\sigma_{11}\sigma_{33})^{1/2}$. Finally, if the third principal axis is along the thickness t of the slab,

$$t'\sigma_m = t(\sigma_{11}\sigma_{22})^{1/2} \qquad (7\text{-}31b)$$

so that three such measurements on two samples satisfying the conditions of Eq. (7-30) provide full information on the conductivity tensor.

If the principal axes of (σ_{ij}) do not lie in the surface, the transformation between the actual dimensions d and t and those of the scaled isotropic medium becomes more complex. In this case we place two axes of a coordinate system along the two surface directions of interest. This coordinate system is related to the principal axis system by a transformation matrix (R_{ij}) of the form of Eq. (3-8). Then a displacement D along one of the two axes, for example (i), has, according to Eq. (3-5), the following components in the principal axis system.

$$(R_{i1}D, R_{i2}D, R_{i3}D)$$

Under the scale transformation Eq. (7-18) this displacement becomes

$$(R_{i1}S_1D, R_{i2}S_2D, R_{i3}S_3D)$$

and therefore the new length D' is related to the old length D along the

i direction by

$$D' = D(\sum_j R_{ij}^2 S_j^2)^{1/2} \tag{7-32}$$

Hence, if a measurement along the i direction in the sample surface yields the product $D'\sigma_m$, the actual dimensions and conductivities involved are given by substitution of Eqs. (7-22), (7-23), and (7-32) as

$$D'\sigma_m = (\sigma_{11}\sigma_{22}\sigma_{33})^{1/2} (\sum_j R_{ij}^2/\sigma_{jj})^{1/2} D \tag{7-33}$$

For the direction normal to the surface we must further take into account that the scale transformation Eq. (7-18) does not preserve angles. The new thickness t' of the scaled sample is given by

$$\frac{\sigma_m}{t'} = (\sum_j \sigma_{jj} R_{ij}^2)^{1/2} \frac{1}{t} \tag{7-34}$$

where R_{ij} refers to the direction cosines of the original normal direction to the surface in the principal axis system.

As expected, Eq. (7-31a) is a special case of Eq. (7-33), as is Eq. (7-31b) of Eq. (7-34).

7.5. ELECTRICAL TRANSPORT IN A MAGNETIC FIELD

Just as a magnetic field **B** exerts a force on a current-carrying wire, we expect that **B** also will affect the details of electrical transport within a material. In fact, these two phenomena are intricately related.

According to the Onsager relations, Eq. (7-10), the electrical conductivity in the presence of **B** obeys the symmetry

$$\sigma_{ij}(\mathbf{B}) = \sigma_{ji}(-\mathbf{B}) \tag{7-35}$$

The conductivity is no longer a symmetric tensor. But we can separate it into a symmetric and an antisymmetric part

$$\sigma_{ij}(\mathbf{B}) = s_{ij}(\mathbf{B}) + a_{ij}(\mathbf{B}) \tag{7-36}$$

with the following separate symmetries.

$$s_{ij}(\mathbf{B}) = s_{ji}(\mathbf{B}) = s_{ji}(-\mathbf{B}) \tag{7-37}$$
$$a_{ij}(\mathbf{B}) = -a_{ji}(\mathbf{B}) = a_{ji}(-\mathbf{B}) \tag{7-38}$$

The field-dependent parts of (s_{ij}) and (a_{ij}) describe all isothermal *galvanomagnetic* phenomena. The two contributions affect electrical conduction in fundamentally different ways. The field-dependent part of (s_{ij}) defines the *magnetoconductivity*, and contributes to the magnitude of anisotropy of the symmetric conductivity tensor. It also enters in the dissipative losses of the material. On the other hand (a_{ij}) exists only in the presence of **B**, and does not participate in heat production. It represents the *generalized Hall effects* in anisotropic matter.

The right-hand equalities of Eqs. (7-36) and (7-37) specify that (s_{ij}) is a function of even powers of **B**, while (a_{ij}) contains only odd powers of **B**. Thus, the first two terms in a power expansion of **B** are

$$s_{ij} = \sigma_{ij} + \sum_{k,l} A_{ijkl} B_k B_l \tag{7-39}$$

$$a_{ij} = \sum_k P_{ijk} B_k \tag{7-40}$$

with the construction of higher terms in the series following the same pattern. Both P_{ijk} and A_{ijkl} have more than two indices and therefore represent tensors of rank higher than two. Because of its particular symmetry, however, P_{ijk} can also be discussed in terms of our formulation of second rank tensors. This is taken up in the next section. Special properties involving the symmetry aspects of A_{ijkl} are part of Chapter 13.

7.6. SYMMETRY OF THE HALL EFFECT

Strictly speaking, the Hall effect is defined by the coefficient inverse to P_{ijk} which arises when the expansion of the resistivity tensor ρ is carried out in the same manner as that for σ in Section 7-5. This inverse coefficient occurs in the relation

$$E_i = \sum_{j,k} R_{ijk} J_j B_k \tag{7-41}$$

Because of Eq. (7-38) both R_{ijk} and P_{ijk} obey the intrinsic symmetry

$$R_{ijk} = -R_{jik}, \qquad P_{ijk} = -P_{jik} \tag{7-42}$$

In addition, of course, both tensors are subject to any restrictions imposed by crystal symmetry. All considerations of crystal symmetry apply equally to R and P. We carry them out explicitly for R_{ijk}.

Formally, R_{ijk} represents a vector–tensor connection, with possibly 27 coefficients. Because of the antisymmetry of the first index pair, however, it can also be thought of as a connection between two axial vectors, the vector product $(\mathbf{E} \times \mathbf{J})$ and $\mathbf{B}$. Therefore, allowing for proper identification of the index pairs with single vector indices, we can take over fully the discussion of symmetry already carried out for the second rank tensor of a vector–vector connection.

The identification of index pairs with single indices is obviously the cyclic one associated with a vector product that was already used in Chapter 4 (see Eq. 4-12)):

$$ij = 23 \quad 31 \quad 12$$
$$(i) = (1) \quad (2) \quad (3)$$
(7-43)

Once the correspondence of Eq. (7-43) is established, we can easily specify the form of the Hall effect tensor $R_{(i)j}$ in any symmetry. For example, in the full tetragonal group it takes the same form as the polarizability tensor α (although it is not intrinsically symmetric) of

$$\begin{pmatrix} R_{(1)1} & 0 & 0 \\ 0 & R_{(1)1} & 0 \\ 0 & 0 & R_{(3)3} \end{pmatrix} \qquad (7\text{-}44)$$

Equation (7-44) defines the following values for R_{ijk} in three-index notation.

$$\begin{array}{lll} R_{(1)1} = R_{231} & R_{(1)2} = R_{232} = 0 & R_{(1)3} = R_{233} = 0 \\ R_{(2)1} = R_{311} = 0 & R_{(2)2} = R_{312} = R_{231} & R_{(2)3} = R_{313} = 0 \quad (7\text{-}45) \\ R_{(3)1} = R_{121} = 0 & R_{(3)2} = R_{122} = 0 & R_{(3)3} = R_{123} \end{array}$$

Therefore the Hall effect relation, Eq. (7-41), which in general requires six terms on each right-hand side, is simplified considerably:

$$E_1 = R_{123}J_2B_3 - R_{231}J_3B_2$$
$$E_2 = R_{123}J_1B_3 + R_{231}J_3B_1 \qquad (7\text{-}46)$$
$$E_3 = R_{231}(J_1B_2 - J_2B_1)$$

Hence two Hall constants characterize a medium of full tetragonal symmetry. In some of the lower groups of tetragonal symmetry a second rank tensor is no longer required to be symmetric. Here the number of independent constants is three, and Eq. (7-43) contains off-diagonal entries. Thus, the Hall effect can distinguish between a greater number of point group symmetries than the dielectric or resistivity tensor. Finally, in a cubic (or isotropic) medium, there is only one constant, and the Hall effect can be written

$$\mathbf{E} = R\mathbf{J} \times \mathbf{B} \tag{7-47}$$

where $R = R_{123} = R_{231} = R_{312}$.

In symmetries involving inversions we must decide whether (R) is polar or axial. Viewed as a second rank tensor, $R_{(i)j}$ connects two axial vectors—a vector product of polar vectors and the axial magnetic field—and is therefore a polar second rank tensor. It does not distinguish between rotations and rotation-inversions. Viewed as a third rank tensor, it connects two polar and one axial vector. Hence it is an axial third rank tensor (i.e., its transformation involves an extra factor (-1) whenever an inversion is involved), so that the result is the same: it does not recognize inversions.

7.7. POTENTIAL DISTRIBUTIONS INCLUDING THE HALL EFFECT

The discussion of Section 7-6 gives the full characterization of the Hall effect in all crystal systems. The combination of constants R_{ijk} measured in any particular configuration depends on both the current distribution, as determined by the boundary conditions for the sample, and the intrinsic anisotropy of the material.

In long, thin, and straight conducting rods the current distribution is fully determined by sample geometry, as has already been discussed in connection with Fig. 7-1. The longitudinal current is uniform and the transverse currents vanish. Then Eq. (7-41) predicts that the Hall effect produces transverse electric fields that depend on the direction of $\mathbf{B}$. In fact, they reverse with the reversal of $\mathbf{B}$, and this property is used to separate these fields from any transverse contributions caused by the anisotropy of the resistivity.

In other sample geometries we must solve for the full potential and current distribution in order to interpret any given measurement. For this treatment it is useful to identify the antisymmetric contribution a_{ij} of Eq. (7-40) as a *Hall conductivity* (σ_{ij}^H), so that to first order in $\mathbf{B}$, $\mathbf{E}$ and $\mathbf{J}$ are related by

$$J_i = \sum_j (\sigma_{ij} + \sigma_{ij}^H) E_j \tag{7-48}$$

The inverse of Eq. (7-48) is

$$E_i = \sum_j (\rho_{ij} + \rho_{ij}^H) J_j \tag{7-49}$$

These equations say that the general form of the Hall fields is similar to that caused by anisotropic conductivity: **E** and **J** are not parallel. However, the Hall effect introduces an essential left-right asymmetry in conduction that distinguishes it from anisotropic conductivity, which is symmetric. Formally, this difference is manifested by the impossibility of finding a principal axis system for Eqs. (7-48) or (7-49).

Even in the presence of (σ_{ij}^H), the potential distribution is determined by Eq. (7-17), the same equation that applies in the absence of **B**. Hence, the Hall effect terms enter only in specifying new boundary conditions. This is brought out clearly if the problem is reformulated for the scaled isotropic material already discussed in Sections 7-3 and 7-4. Applying the transformation Eq. (7-18) to Eq. (7-48) results in the relation

$$J_i' = \sum_j [\sigma_m \delta_{ij} + (\sigma_{ij}^H/S_i S_j)] E_j' \qquad (7\text{-}50)$$

indicating that the scaled isotropic medium remains anisotropic in an applied magnetic field. If a surface defined by a normal (n_1', n_2', n_3') bounds the current, then we require at the surface that $J_n' = 0$, or

$$\sum_i J_i' n_i' = 0 \qquad (7\text{-}51)$$

Equation (7-51) specifies the required relations between the electric field components at the surface if J_i' is taken from Eq. (7-50).

In some problems Eq. (7-51) does not have to be used because the potential is specified on all boundaries. Thus, in the geometry of the Corbino disk, shown in Fig. 7-5, **B** is perpendicular to the plane of a circular disk, and two circular ring electrodes specify a potential drop V. If the medium is isotropic, all equipotentials are circular, and the two-dimensional solution of $\nabla^2 \Phi = 0$ in polar coordinates is

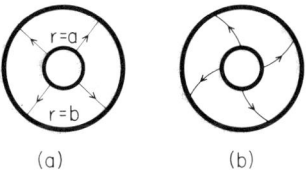

(a)   (b)

Fig. 7-5 The Corbino disk. (a) Current and field lines for **B** = 0. (b) Current lines for **B** normal to the disk. Field lines remain as in (a).

$$\Phi(r, \phi) = V \frac{\ln(r/a)}{\ln(b/a)} \tag{7-52}$$

The Hall conductivity leads to the relations between currents and fields

$$J_1 = \sigma E_1 + \sigma^H E_2, \quad J_2 = -\sigma^H E_1 + \sigma E_2$$

or in polar coordinates

$$J_r = \sigma E(r), \quad J_\phi = -\sigma^H E(r)$$

Hence the current flow lines are determined by the ratio

$$J_\phi/J_r = r d\phi/dr = -\sigma^H/\sigma \tag{7-53}$$

leading to the equation for the flow lines

$$\phi = (-\sigma^H/\sigma) \ln r + \text{constant} \tag{7-54}$$

Equation (7-54) describes a logarithmic spiral such that the current everywhere flows at the same angle with respect to the circular equipotentials.

7.8 TRANSPORT IN MAGNETIC MATERIALS

Crystals belonging to one of the magnetic groups are characterized by a spontaneous magnetization $\mathbf{M}_0$, or by built-in directions of antiferromagnetic ordering that have the same symmetry under space and time inversion as $\mathbf{M}_0$. In all such crystals, the Onsager relations take the extended form described by Eq. (7-10), but applying to both $\mathbf{B}$ and $\mathbf{M}_0$:

$$\sigma_{ij}(\mathbf{M}_0, \mathbf{B}) = \sigma_{ji}(-\mathbf{M}_0, -\mathbf{B}) \tag{7-55}$$

To first power in $\mathbf{B}$, the expansion of Eq. (7-55) is

$$\sigma_{ij}(\mathbf{M}_0, \mathbf{B}) = \sigma_{ij}(\mathbf{M}_0) + \sum_k P_{ijk}(\mathbf{M}_0) B_k \tag{7-56}$$

and according to Eq. (7-55) the symmetry of the $\mathbf{M}_0$-dependent coefficients is

$$\sigma_{ij}(\mathbf{M}_0) = \sigma_{ji}(-\mathbf{M}_0), \quad P_{ijk}(\mathbf{M}_0) = -P_{jik}(-\mathbf{M}_0) \tag{7-57}$$

As a consequence of Eq. (7-57), both $\sigma_{ij}(\mathbf{M}_0)$ and $P_{ijk}(\mathbf{M}_0)$ now contain contributions either symmetric or antisymmetric in the index pair (i, j), so that the analysis carried out in Section 7-5 for $\sigma_{ij}(\mathbf{B})$ can be applied once more, leading, for instance, to equations corresponding to Eqs. (7-37) and (7-38), with $\mathbf{M}_0$ replacing $\mathbf{B}$. For example, if an expansion in powers of $\mathbf{M}_0$ is appropriate, we obtain the relations

$$\sigma_{ij}(\mathbf{M}_0) = \sigma_{ij}^o + \sum_k P_{ijk}^e M_{0k} \qquad (7\text{-}58)$$

with

$$\sigma_{ij}^o = \sigma_{ji}^o, \qquad P_{ijk}^e = -P_{jik}^e \qquad (7\text{-}59)$$

and

$$P_{ijk}(\mathbf{M}_0) = P_{ijk}^o + \sum_l A_{ijkl}^e M_{0l} \qquad (7\text{-}60)$$

with

$$P_{ijk}^o = -P_{jik}^o, \qquad A_{ijkl}^e = A_{jikl}^e \qquad (7\text{-}61)$$

The new terms in these expansions have been designated by the symbols appropriate to Hall effect and magnetoresistance because they show behavior very similar to these phenomena. To distinguish them, we speak of *ordinary* and *extraordinary* effects, as labeled by the superscripts o and e.

Equations (7-58) and (7-60) give rise to phenomena characteristic of magnetic symmetry, such as a Hall effect independent of $\mathbf{B}$, and a magnetoresistance linear in $\mathbf{B}$. The restrictions imposed by crystal symmetry on the extraordinary coefficients are obtained by the same treatment that applies to the ordinary galvanomagnetic coefficients. With respect to (A_{ijkl}^e) we must take into account, however, that the indices k and l are not necessarily interchangeable.

If $\mathbf{M}_0$ does not exist, but we have a magnetic symmetry, the foregoing treatment can still be carried out formally. But it is usually preferable to specify directly the transformation properties of σ_{ij}, P_{ijk}, etc. under the time reversal operation $\mathcal{R}$ discussed in Chapter 6. They are given by

$$\mathcal{R}(\sigma_{ij}) = \sigma_{ji} \qquad (7\text{-}62)$$

$$\mathcal{R}(P_{ijk}) = -P_{jik} \qquad (7\text{-}63)$$

The same rule applies if $\mathcal{R}$ exists in combination with a spatial symmetry element.

Problems

7-1. Determine in which of the following symmetries the observation of the equality $\sigma_{ij} = \sigma_{ji}$ is a nontrivial verification of the Onsager relations: (m), (mm), $(\bar{3})$, $(\bar{6}m2)$.

7-2. In a conducting crystal of symmetry $\bar{4}2m$, find
   (a) the angle ψ between $\mathbf{J}$ and $\mathbf{E}$ for a given direction of $\mathbf{E}$;
   (b) the direction of $\mathbf{E}$ that maximizes ψ;
   (c) the maximum value of ψ.

7-3. The rate of Joule heat generation per unit volume accompanying electric current flow is given by $\mathbf{E} \cdot \mathbf{J}$. Using the equations of Section 7-1, show that
   (a) the conventional conductivity σ_{ij} is related to the generalized conductivity L_{ij} by

$$L_{ij} = T\sigma_{ij}$$

   where T is the absolute temperature.
   (b) The components of conductivity obey the inequality

$$\sigma_{ii}\sigma_{jj} > \sigma_{ij}^2$$

7-4. If the sample of Fig. 7-1 carries a current density of 10^5 amp/m^2 and is characterized by $\rho_{11} = 10^{-8}$ ohm-m, $\rho_{12} = 2 \cdot 10^{-9}$ ohm-m, $\rho_{13} = 3 \cdot 10^{-9}$ ohm-m, determine the electrostatic surface charge density on the lateral faces.

7-5. Carry out the qualitative graphical construction of equipotentials and current flow lines, analogous to that for the long thin rod in Section 7-2, if the conducting sample consists of a thin anisotropic slab between closely spaced parallel highly conducting planes.

7-6. Show that the requirement of continuity of current under the transformation Eq. (7-18) is given by

$$J_1' = \frac{1}{S_2 S_3} J_1, \text{ etc.}$$

and that this relation is satisfied, using Eqs. (7-19) and (7-20).

7-7. Use the method of Section 7-3 to derive the solution for constant current flow given by Eq. (7-14).

7-8. Assuming that Fig. 7-2 represents two-dimensional flow (the distribution of $\mathbf{E}$ and $\mathbf{J}$ is the same in all planes $z = $ constant) in a slab of thickness t, carry out the transformations to give the corresponding sample shape in the scaled isotropic medium. Take the rod axis to lie along a principal direction, with $\rho_{22} = 0.5\rho_{11}$.

7-9. Show that Eq. (7-33) can also be written as

$$D'\sigma_m = (\sigma_{11}\sigma_{22}\sigma_{33})^{1/2} (\rho_{ii}^s)^{1/2}$$

Electrical Conduction

where ρ_{ii}^s is a diagonal component of the resistivity in the system of coordinates in which the conductance measurements are carried out.

7-10. In a rectangular arrangement of four probes on the same face of a conducting slab, the current flows between two probes a distance a apart, and the potential drop is measured between two probes spaced by a along a line a distance b from the line of current probes. The voltage–current relationships in such an arrangement are, for an isotropic conductor of resistivity ρ

$$t \gg a, b \qquad V = \frac{\rho}{\pi} \frac{I}{b} [1 - \frac{1}{(1 + a^2/b^2)^{1/2}}]$$

$$t \ll a, b \qquad V = \frac{\rho}{2\pi} \frac{I}{t} \ln(1 + \frac{a^2}{b^2})$$

where t is the slab thickness.

Use the method of Section 7-4 to determine the corresponding formulas in an anisotropic medium. Assume that the edges of the rectangle line up with two principal directions of (ρ_{ij}).

7-11. Verify Eq. (7-34).

7-12. Use Eqs. (7-37) and (7-39) to establish the symmetry of the quadratic magnetoconductivity coefficients

$$A_{ijkl} = A_{jikl} = A_{jilk}$$

and show that to second power in **B** Joule heat dissipation only involves the tensor (A). What restrictions are imposed on (A) by the requirement Eq. (7-3)?

7-13. Use the method of Section 7-6 to determine the Hall field in a crystal of symmetry $(4/m)$.

7-14. How many Hall constants are required in the symmetry groups $(\bar{1})$, $(2/m)$, (3), $(6mm)$, $(m3m)$?

7-15. Use the methods of Chapter 4 to determine the Hall constant of polycrystalline material by considering it as (a) a second rank polar tensor; (b) a third rank axial tensor.

7-16. Determine the structure of the resistivities and Hall constants and of their inverse coefficients in the crystal system $(3m)$. Write out the relation between (ρ_{ij}) and (σ_{ij}) and between (R_{ijk}) and (P_{ijk}).

7-17. Express the boundary conditions on Φ following from Eq. (7-51) for current flowing in a long rod of square cross section along the x direction if the medium has the symmetry $(3m)$ of Problem 7-16.

7-18. (a) Calculate the total angular and radial currents I_ϕ, I_r in the Corbino disk of Section 7-7 and Fig. 7-5.
(b) Show that if the material is slightly anisotropic, as measured by a small parameter $\delta = (\sigma_{22} - \sigma_{11})/\sigma_{11}$, then, to first order in δ, I_ϕ remains unchanged, while σI_r becomes $\sigma_{11} I_r (1 + \delta/2)$.

7-19. Show that the linear magnetoresistance of Eq. (7-60) contributes to the Joule heat dissipation, while the extraordinary Hall effect of Eq. (7-58) does not. Generalize this result to show that dissipative effects are even under time reversal $\Re$, while lossless effects reverse under this operation.

7-20. Show that the antisymmetric part of the magnetization-dependent conductivity

$$\frac{1}{2}(\sigma_{ij}(\mathbf{M}_0) - \sigma_{ji}(\mathbf{M}_0))$$

transforms like an axial vector that reverses under $\Re$, and that it therefore exists in all magnetic groups permitting a spontaneous magnetization.

7-21. Devise an experimental means for separating the ordinary and extraordinary Hall effects in a ferromagnetic material.

Bibliography

J. F. Nye, *Physical Properties of Crystals*, Oxford Univ. Press, London and New York (1957), Chapter 11.

S. Bhagavantam, *Crystal Symmetry and Physical Properties*, Academic Press, New York (1966), Chapter 17.

W. P. Mason, *Crystal Physics of Interaction Processes*, Academic Press, New York (1966), Chapter 9.

CHAPTER 8

Thermoelectricity

The transport of electrical charge treated in Chapter 7 is always accompanied by both transport and dissipation of electrical energy. An electric current develops temperature gradients and thermal currents whose steady-state distribution depends on the thermal properties and on the thermal boundary conditions of the material. If the electrical and thermal properties are temperature-dependent, as is usually the case, it is therefore necessary to obtain a self-consistent solution involving both electrical and thermal conduction.

Beyond this obvious interconnection, electrical and thermal flows are further related because they represent two simultaneous nonequilibrium processes in the same medium. The Onsager relations of Eq. (7-10) fully apply to such a setting of two or more different transport phenomena as long as the forces and fluxes of each process are properly defined. Obviously, since in the case of simultaneous electrical and thermal flow there are six independent components of current, the two indices in L_{ij} then run from 1 to 6. Therefore (L_{ij}) is a second rank tensor in six-dimensional space, and the linear relation between the currents and the driving forces is given by

$$J_i = \sum_j L_{ij} X_j, \quad i,j = 1, ..., 6 \qquad (8\text{-}1)$$

Equation (8-1) predicts that some of the off-diagonal elements of (L_{ij}) connect thermal currents to electrical forces, or electrical currents to thermal forces. These off-diagonal elements of (L_{ij}) describe the *thermoelectric effects*. This chapter treats thermoelectric phenomena in crystals, starting with the unifying approach of irreversible thermodynamics express by Eq. (8-1). To give meaning to this equation we must first identify the proper flux–force pairs J_i, X_i for which it is valid. Thereafter it is important to develop the effect of crystal symmetry on the structure of (L_{ij}). Finally, we discuss various formulations of thermoelectric phenomena applicable for specific electrical and thermal boundary conditions. Some of these can be identified with the classical thermoelectric effects in isotropic matter. Others only exist in anisotropic crystals.

8.1. CURRENTS AND DRIVING FORCES

Let us consider unit volume of a material traversed by electrical and thermal current densities $\mathbf{J}$ and $\mathbf{J}_q$. The rate of change in time of the energy density U of this volume element is

$$\frac{\partial U}{\partial t} = -\nabla \cdot (\overline{\Phi} \mathbf{J}) - \nabla \cdot \mathbf{J}_q \tag{8-2}$$

where $\overline{\Phi}$ is the local *electrochemical potential*, a generalization of the electrical potential that takes into account the local chemical composition of the material. This generalization is important here because some of the interesting thermoelectric effects occur at the boundary between different materials.

In addition to Eq. (8-2), U obeys the conservation law

$$\frac{\partial U}{\partial t} + \nabla \cdot \mathbf{J}_u = 0 \tag{8-3}$$

where $\mathbf{J}_u$ is the current of energy. Comparing Eqs. (8-2) and (8-3), the energy current is related to the other currents by

$$\mathbf{J}_u = \overline{\Phi} \mathbf{J} + \mathbf{J}_q \tag{8-4}$$

The entropy density S is related to U through the first law of thermodynamics. With the thermodynamic work only arising from the transport of charge Q, this law leads to the time rates of change

$$T \frac{\partial S}{\partial t} = \frac{\partial U}{\partial t} - \overline{\Phi} \frac{\partial Q}{\partial t} \tag{8-5}$$

and by using Eqs. (8-3) and (8-4), and the law of conservation of charge, Eq. (8-5) can be rewritten as

$$\frac{\partial S}{\partial t} = -\frac{1}{T} \nabla \cdot \mathbf{J}_q - \frac{1}{T} \mathbf{J} \cdot \nabla \overline{\Phi}$$

Recasting the first term on the right side to have a form similar to the second term, we finally obtain

$$\frac{\partial S}{\partial t} + \nabla \cdot \left(\frac{\mathbf{J}_q}{T}\right) = -\mathbf{J}_q \cdot \frac{\nabla T}{T^2} - \mathbf{J} \cdot \frac{\nabla \overline{\Phi}}{T} \tag{8-6}$$

Equation (8-6) would represent a conservation law of entropy if its right-hand side were equal to zero. Since the processes under consideration are not reversible, there is an increase in entropy, and the rate of entropy production must be given by the right side of Eq. (8-6)

$$\frac{dS}{dt} = -\mathbf{J}_q \cdot \frac{\nabla T}{T^2} - \mathbf{J} \cdot \frac{\nabla \overline{\Phi}}{T} \tag{8-7}$$

As established in Section 7-1, the form of Eq. (8-7) permits writing down directly the proper linear relations to which the Onsager relations apply. Rather than cast them into the generalized form established by Eq. (8-1), let us immediately write these linear relations in the more practical form directly related to the traditional coefficients. If the material is *isotropic*, Eq. (8-7) requires the two laws of irreversible flow

$$\mathbf{J} = -\sigma T (\nabla \overline{\Phi}/T) + \beta T^2 (\nabla T/T^2) \tag{8-8}$$

$$\mathbf{J}_q = \beta' T (\nabla \overline{\Phi}/T) - K' T^2 (\nabla T/T^2) \tag{8-9}$$

The symbols have been chosen so that σ is the traditional *isothermal* electrical conductivity, and K' is the *field-free* thermal conductivity. The coefficients β and β' describe the cross coupling between thermal and electrical effects.

In anisotropic matter, the four scalars $\sigma, K', \beta, \beta'$ of Eqs. (8-8) and (8-9) are replaced by four second rank tensors. These tensors obey the symmetry relations imposed by the Onsager relations Eq. (7-10) as they apply to Eq. (8-1)

$$\sigma_{ij}(\mathbf{B}) = \sigma_{ji}(-\mathbf{B}), \quad K'_{ij}(\mathbf{B}) = K'_{ji}(-\mathbf{B}), \quad T\beta_{ij}(\mathbf{B}) = \beta'_{ji}(-\mathbf{B}) \tag{8-10}$$

Hence the Onsager relations specify the intrinsic symmetry of the electrical and thermal conductivities, and they establish an interdependence of the thermoelectric tensors β and β'.

For practical purposes it is convenient to resolve Eqs. (8-8) and (8-9) and define a new set of coefficients

$$-\nabla \overline{\Phi} = \rho \mathbf{J} + Q \nabla T \tag{8-11}$$

$$\mathbf{J}_q = TP\mathbf{J} - K \nabla T \tag{8-12}$$

Both equations are now a mixture of currents and forces on the right side, but it can be shown that the new sets of coefficients also obey the Onsager relations

$$\rho_{ij}(\mathbf{B}) = \rho_{ji}(-\mathbf{B}), \quad K_{ij}(\mathbf{B}) = K_{ji}(-\mathbf{B}), \quad P_{ij}(\mathbf{B}) = Q_{ji}(-\mathbf{B}) \tag{8-13}$$

Equations (8-11) and 8-12) are more useful than the earlier Eqs. (8-8) and (8-9) because their right-hand sides refer directly to the electrical current and the temperature gradient, both quantities that are under easy control. ρ is the isothermal resistivity and K is the thermal conductivity in the absence of electrical current. The coefficient Q defines the electrical potential drop accompanying a difference in temperatures in an open circuit, while P specifies the thermal current accompanying an isothermal electrical current. Equation (8-13) indicates that these last two processes are interrelated.

8.2. THERMOELECTRIC HEAT GENERATION

The steady state of simultaneous current and heat flows is described by the conditions

$$\nabla \cdot \mathbf{J}_u = 0, \quad \nabla \cdot \mathbf{J} = 0 \tag{8-14}$$

Applying Eq. (8-14) to Eq. (8-4), and eliminating $\nabla \Phi$ and $\mathbf{J}_q$ through Eqs. (8-11) and (8-12), we obtain, as a description of this steady state, the equation

$$-\nabla \cdot (K \nabla T) = \mathbf{J} \cdot (\rho \mathbf{J}) - \nabla \cdot (TP\mathbf{J}) + \mathbf{J} \cdot (Q \nabla T) \tag{8-15}$$

Equation (8-15) is the differential equation for the steady-state temperature distribution in the presence of an electrical current. For small constants P and Q all terms on the right-hand side can be considered sources of heat production in unit volume. The first term gives the Joule heat, and the other two describe the thermoelectric sources of heat generation. Equation (8-15) also applies fully in anisotropic material if the products inside the brackets are carefully identified as products of tensors and vectors. This is made clear in the component notation developed later. First, however, it is interesting to carry the symbolic notation of Eq. (8-15) one step further. Apart from the Joule heat, the thermoelectric heat sources of Eq. (8-15) can be transformed to give a rate of heat production

$$dq/dt = -T \nabla \cdot (P\mathbf{J}) + \mathbf{J} \cdot ((Q - \widetilde{P}) \nabla T) \tag{8-16}$$

In isotropic material, or material that requires Q and P to be symmetric tensors, the last term on the right side vanishes. Hence all the normal thermoelectric effects are contained in the first term. The last term is an effect peculiar to sufficiently anisotropic materials. Because of Eq. (8-13) it can only occur in non-vanishing magnetic fields. Apparently it has never been reported, although, as shown later, it should be observable.

The thermoelectric coefficients P and Q are second rank tensors without intrinsic symmetry. Hence in a given crystal, they exhibit the symmetry characteristic of a general second rank in that medium, following the rules developed in Chapters 3 and 4. Furthermore, Eq. (8-13) indicates that the elements of P are expressible in terms of those of Q, and vice versa, so that thermoelectric effects are fully specified if either P or Q is known. As discussed in the following sections, however, it is convenient to keep both coefficients because traditional thermoelectric phenomena directly refer to one or the other. As a result, we expect that there must be relations between these thermoelectric effects.

In the discussion of the various thermoelectric effects it is assumed that the matter constants P and Q are functions of both position x_i and temperature T. The dependence on position relates to the change of composition in the conducting material, which may occur gradually or be abrupt, as at a junction between two different materials. Furthermore, the components of $\mathbf{J}$ may also be position dependent, subject to the requirement of continuity.

Taking into account these functional dependences, we have, as the normal thermoelectric heat production of Eq. (8-16) given in component notation,

$$\frac{dq}{dt} = -\sum_{i,j} T \frac{\partial P_{ij}}{\partial x_i} J_j - \sum_{i,j} T \frac{\partial P_{ij}}{\partial T} \frac{\partial T}{\partial x_i} J_j - \sum_{i,j} TP_{ij} \frac{\partial J_j}{\partial x_i} \qquad (8\text{-}17)$$

while the last term of Eq. (8-16) takes the form

$$-\sum_{ij} (P_{ij} - Q_{ji}) \frac{\partial T}{\partial x_i} J_j \qquad (8\text{-}18)$$

Formally, it has the same structure as the second term in Eq. (8-17). Hence Eqs. (8-17) and (8-18) give rise to three groups of characteristic effects. These are discussed separately in the next section.

8.3. THERMOELECTRIC EFFECTS

The three terms of Eq. (8-17) are designated, respectively, the Peltier, Thomson, and Bridgman effects. Each refers to a specific experimental situation.

A. Peltier Effects

The Peltier effects are described by the first sum in Eq. (8-17). They predict local heat generation or absorption when an electric current flows at constant temperature through a material whose properties are not uniform. The sign of the heat generation changes as the direction of $\mathbf{J}$ is reversed. One expects a

Peltier effect in any region of a material where, for example, composition or strain is changing. The classical application occurs, of course, in the region of abrupt change of properties where two uniform media of differing properties are joined.

At such a junction the heat generated in an element of interface dA is given by

$$-T\,d\mathbf{A} \cdot (\Delta P \mathbf{J}) \tag{8-19}$$

where ΔP is the difference in P of the two materials. The components of $\mathbf{J}$ are assumed to be continuous across the junction. Applied to a plane interface of area A with a normal $\mathbf{n}$, Eq. (8-19) becomes

$$dq/dt = -TA \sum_{i,j} n_i \Delta P_{ij} J_j \tag{8-20}$$

where $\Delta P_{ij} = P_{ij}(2) - P_{ij}(1)$ when $\mathbf{n}$ points from medium 1 into medium 2. The components of ΔP_{ij} refer to a coordinate system common to both materials, and since in general this system will not coincide with the crystal symmetry of either material, all nine components of (ΔP_{ij}) are nonvanishing. In that case the most convenient coordinate system is one whose axes coincide with $\mathbf{n}$ or $\mathbf{J}$.

As an example, let us determine the Peltier heat generation in the junction shown in Fig. 8-1.

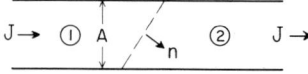

Fig. 8-1 Two-dimensional Peltier junction. A constant current density $\mathbf{J}$ traverses a planar interface with normal $\mathbf{n}$. For unit depth the area of the cross section of the two conductors is A.

If $\mathbf{J} = (J, 0, 0)$ and $\mathbf{n} = (\cos\alpha, -\sin\alpha, 0)$, then from Eq. (8-20).

$$\begin{aligned}\frac{dq}{dt} &= -T\frac{A}{\cos\alpha}\,[\,\cos\alpha\,\Delta P_{11} - \sin\alpha\,\Delta P_{21}\,]\,J \\ &= -TA\,[\,\Delta P_{11} - \tan\alpha\,\Delta P_{21}\,]\,J\end{aligned} \tag{8-21}$$

The first term in Eq. (8-21) has the expected form. The second term, on the other hand, seems to imply that the heat generation at the junction depends on the details of the geometry of the contact between the two materials, contrary to experience.

This contradiction is resolved by taking into account that Eq. (8-20) also applies to the lateral boundary between each conductor and vacuum. From Fig. 8-2, we see that $\mathbf{n}_t = (0, 1, 0)$, and assuming that in vacuum $(P_{ij}) = 0$, Eq. (8-20)

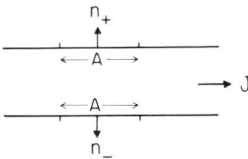

Fig. 8-2 Lateral Peltier junction in anisotropic material between the conductor and vacuum.

predicts a heat generation per unit area

$$dq/dt = TP_{21}J \qquad (8\text{-}22)$$

at the upper surface. At the lower surface the same result holds with opposite sign. If the junction is inclined, as in Fig. 8-1, an additional area $A \tan \alpha$ is exposed on the upper surface to medium 1 rather than medium 2. As a result of these lateral junctions, there is a net change in lateral heat generation given by

$$A \tan \alpha \, [TP_{21}(1)J - TP_{21}(2)J] = -A \tan \alpha \, \Delta P_{21} J$$

that exactly counterbalances the second term in Eq. (8-21). It is obvious that the generalization of this result is that *the total thermoelectric heat generation at a junction is independent of the junction geometry as long as the region of the junction is properly defined.* This result remains true if because of a strong difference in electrical properties the current flow in the neighborhood of the junction is distorted. In that case the Bridgman terms provide the necessary balance.

Equation (8-22) is of interest in its own right because it predicts that if the anisotropic conductor remains isothermal, it can act as a lateral heat pump, for example, extracting heat from the lower medium in Fig. 8-2 and transferring it to the upper medium. Except in isotropic media such lateral pumping is always possible if $\mathbf{J}$ flows in other than directions of symmetry, because in the coordinate system tied to $\mathbf{J}$, as developed in Eqs. (8-21) and (8-22), (P_{ij}) will have its full complement of nine components and thus will include the necessary off-diagonal terms.

B. Thomson Effects

The Thomson effects refer to the second term in Eq. (8-17). There is additional heat generation per unit volume if **J** flows in the presence of a temperature gradient. This effect exists in isotropic materials, and the generalization to anisotropic media merely adds the possibility that **J** and ∇T need not be parallel. The sign of dq/dt reverses with the reversal of either **J** or ∇T.

The additional effect described by Eq. (8-18) has no isotropic counterpart. It even vanishes in anisotropic matter as long as **B** = 0. But in a magnetic field it contributes terms odd in **B**.

C. Bridgman Effects

The last term in Eq. (8-17) refers to local heating or cooling when an electric current changes direction in a crystal. It is nonvanishing only if (P_{ij}) is not a scalar, since $\nabla \cdot \mathbf{J} = 0$. It refers to uniform material, but as discussed earlier under the Peltier effects, it enters into the heat produced at a junction if the electrical properties of the two media force a distortion of current flow near the junction.

D. Seebeck Effects

These describe the relation between the electrochemical potential $\overline{\Phi}$ and the temperature T in the absence of electrical current. It follows from Eq. (8-11) that the two gradients are connected by

$$-\frac{\partial \overline{\Phi}}{\partial x_i} = \sum_j Q_{ij} \frac{\partial T}{\partial x_j} \qquad (8\text{-}23)$$

where Q_{ij} may be position and temperature dependent. Hence the two gradients are not necessarily parallel, and a longitudinal temperature gradient sets up transverse voltage drops. Considering that in thin rods carrying uniform thermal current, ∇T also has transverse components, the measurement of (Q_{ij}) is not simple.

In a thermocouple this difficulty is avoided by defining a total temperature difference between the two probes for measuring the potential. As shown in Fig. 8-3 the open circuit voltage $V = \overline{\Phi}_2 - \overline{\Phi}_1$ is measured in an isothermal environment, between leads of the same material and properties, so that the chemical part of $\overline{\Phi}$ is the same on both sides of the break. By integrating Eq. (8-23) around the circuit of Fig. 8-3 for an infinitesimal temperature difference dT, we obtain

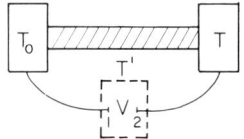

Fig. 8-3 Thermocouple circuit.

$$dV/dT = -(Q_{11} - Q) = \Delta Q \qquad (8\text{-}24)$$

where the crystal rod axis is assumed to be along the x axis and Q refers to the thermoelectric property of the isotropic potential leads. ΔQ is defined to conform with the definition of ΔP given earlier. By selecting rod axes in sufficiently different directions with respect to crystal axes, all the components of (Q_{ij}) can be determined.

8.4. THERMOELECTRIC RELATIONS

The four effects discussed in Section 8-3 are all described by the single tensor (P_{ij}), and if, as originally supposed, these different thermoelectric phenomena can be treated separately and independently, then there must be relations between them.

Thus, in isotropic matter, we have the traditional coefficients S, Π, τ defined by

Thermoelectric power of thermocouple junction: $\qquad S$
Peltier heat *absorbed* at a unit area junction: $\qquad \Pi J$
Thomson heat per unit temperature gradient
  supplied to maintain the temperature distribution: $\qquad \tau J$

These coefficients are related to those of Section 8-3 by

$$S = \Delta Q, \qquad \Pi = T \Delta P, \qquad \Delta\tau = T\frac{d\,\Delta P}{dT} \qquad (8\text{-}25)$$

and as a result of Eq. (8-13) there exist the relations

$$\Pi(\mathbf{B}) = TS(-\mathbf{B}), \qquad \Delta\tau(\mathbf{B}) = T\frac{dS}{dT}(-\mathbf{B}) \qquad (8\text{-}26)$$

In anisotropic media, we could define the corresponding tensor coefficients S_{ij}, Π_{ij}, $\Delta\tau_{ij}$. All of these coefficients refer to two media. But because both media are separately anisotropic, it is clear that relations of the form of Eq. (8-26) cannot in general be established unless the experimental conditions are defined much more carefully than in the isotropic case. In any given situation, the relations that apply can be easily derived from the formulas given here.

This is illustrated by the following specific example. Consider a long Bismuth crystal parallel to the threefold z axis. A temperature gradient exists along z, current flows along z, and junctions are made to an isotropic conductor on the z faces. Then we obtain the following three expressions.

Thermoelectric power:

$$\frac{dV}{dT} = -(Q_{33} - Q_{\text{iso}})$$

Peltier heating at unit area junction where unit current flows out of the crystal:

$$\frac{dq}{dt} = -T(P_{\text{iso}} - P_{33})$$

Thomson heat per unit volume, current, and temperature gradient:

$$\frac{dq}{dt} = -[T\frac{\partial P_{33}}{\partial T} + (P_{33} - Q_{33})]$$

If we use the definitions of Eq. (8-25), and remember that Δ is defined, for example, by $(P_{\text{iso}} - P_{\text{crystal}})$, then we obtain the thermoelectric relations for this experimental arrangement

$$\Pi(\mathbf{B}) = TS(-\mathbf{B}), \qquad \Delta\tau(\mathbf{B}) = T\frac{dS}{dT}(-\mathbf{B}) - (S(\mathbf{B}) - S(-\mathbf{B}))$$

This particular generalization is interesting because in Bismuth $S(\mathbf{B})$ and $S(-\mathbf{B})$ differ for certain directions of $\mathbf{B}$, as is discussed in Problems 8-7 and 8-9.

Problems

8-1. Show that the coefficients of Eq. (8-8) and (8-9) must obey the inequalities

$$\sigma > 0, \qquad \kappa' > 0, \qquad \sigma\kappa' > \beta\beta'$$

Thermoelectricity   101

8-2. Show that the coefficients of Eqs. (8-11) and (8-12) are related to those of Eqs. (8-8) and (8-9) by

$$\rho = \sigma^{-1} \qquad TP = -\beta'\sigma^{-1}$$

$$K = K' - \beta'\sigma^{-1}\beta, \qquad Q = -\sigma^{-1}\beta$$

and verify that if Eq. (8-10) is obeyed, Eq. (8-13) follows.

8-3. Generalize the results of Problem 8-1 if all coefficients are second rank tensors.

8-4. If **B** = 0, determine the number of independent constants, and the structure of (L_{ij}) of Eq. (8-1), in the crystal symmetries $(2/m)$; $(\bar{3}m)$; (6); (23).

8-5. Given a general second rank tensor T_{ij} which is a function of a vector **B**: $T_{ij}(B_1, B_2, B_3) = T_{ij}(B_k)$. Show that if the transformation (R_{ij}) is a symmetry operation, the tensor components are related by the set of linear equations

$$T_{ij}\left(\sum_l R_{ml}B_l\right) = \sum_{k,l} R_{ik}R_{jl}T_{kl}(B_m)$$

8-6. Given a crystal of symmetry (2), with the twofold axis along z. Using Problem 8-5 show the following.
(a) If **B** is along the z axis, (P_{ij}) has the form

$$(P_{ij}) = \begin{pmatrix} P_{11} & P_{12} & 0 \\ P_{21} & P_{22} & 0 \\ 0 & 0 & P_{33} \end{pmatrix}$$

with $P_{ij}(\mathbf{B})$ an arbitrary function of **B**.
(b) If **B** is normal to the z axis, the foregoing components also exist, but with the restriction $P_{ij}(\mathbf{B}) = P_{ij}(-\mathbf{B})$. In addition, (P_{ij}) allows the components $P_{13}, P_{23}, P_{31}, P_{32}$ with the restriction $P_{ij}(\mathbf{B}) = -P_{ij}(-\mathbf{B})$.

8-7. (a) Apply the rule of Problem 8-5 to show that if in a crystal of Bismuth (symmetry $(\bar{3}m)$) **B** is in the plane normal to the threefold axis and bisecting two binary axes, then

$$Q_{33}(\mathbf{B}) = Q_{33}(-\mathbf{B})$$

but that if **B** is along a binary axis, this equality is not required.
(b) Does the same argument hold in the symmetry (32)?

8-8. Use the results of Problem 8-6 to determine the number of components of the tensor $(P_{ij} - Q_{ji})$ of Eq. (8-18) in the system (2) for **B** either parallel or perpendicular to the twofold axis.

8-9. If an isotropic material is subjected to a magnetic field **B**, then the coordinate system can be chosen for **B** to lie along the z axis.
(a) Show that in this coordinate system the material has a thermoelectric tensor (P_{ij}) given by

$$\begin{pmatrix} P_{11} & P_{12} & 0 \\ -P_{12} & P_{11} & 0 \\ 0 & 0 & P_{33} \end{pmatrix}$$

where $P_{11}(\mathbf{B}) = P_{11}(-\mathbf{B})$, $P_{33}(\mathbf{B}) = P_{33}(-\mathbf{B})$, and $P_{12}(\mathbf{B}) = -P_{12}(-\mathbf{B})$.

(b) Use the Onsager relations to show that the tensor $(P_{ij} - Q_{ji})$ vanishes in this material.

8-10. Show that in groups of magnetic symmetry the thermoelectric coefficients may differ from those of the corresponding nonmagnetic group only in their dependence on $\mathbf{B}$.

8-11. Figure 8-4 shows a crystal shaped as an arc of a ring, making contact with two different isotropic materials. A uniform current density flows as indicated, and the thermoelectric tensors of the three materials are $P^{(a)}$, P_{ij}, $P^{(b)}$ respectively.

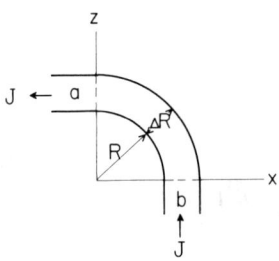

Fig. 8-4 Geometry of a quarter-circle ring of anisotropic material discussed in Problem 8-11.

(a) Calculate the net Peltier heat generated at the four lateral faces of the ring, and at the two junctions.
(b) Calculate the Bridgman heat generated in the ring.
(c) Show that if the full arrangement of Fig. 8-4 is considered as a single junction at temperature T, the heat generated at the junction is independent of the properties of the ring portion.

8-12. Figure 8-5 shows current flowing through a rectangular bend in a crystal described by

$$(P_{ij}) = \begin{pmatrix} P_{11} & 0 & 0 \\ 0 & P_{22} & 0 \\ 0 & 0 & P_{33} \end{pmatrix}$$

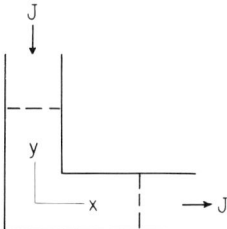

Fig. 8-5 Current flowing in a rectangular crystal bend for the geometry of Problem 8-12.

(a) Show that despite nonuniform current flow, there is no heat production at the lateral faces.
(b) Use Eq. (8-16) to calculate the heat generated in the crystal by applying Gauss's Law, thus avoiding the need to determine the current distribution. Show that this is independent of the location of the junctions as long as they are traversed by a uniform current density.
(c) Verify that if the crystal bend is terminated by the same isotropic material on both ends, the heat generation of the overall junction is zero.

Bibliography

J. F. Nye, *Physical Properties of Crystals*, Oxford Univ. Press, London and New York (1957), Chapter 12.

L. D. Landau and E. M. Lifshitz, *Electrodynamics of Continuous Media*, Addison-Wesley, Inc., Reading, Mass. (1960), Chapter 3, Section 25.

R. Becker, *Theory of Heat*, Springer, New York (1967), Chapter 7.

P. W. Bridgman, *The Thermodynamics of Electrical Phenomena in Metals*, Macmillan, New York (1934).

CHAPTER 9

Crystal Optics

Crystal optics deals with the propagation of electromagnetic waves in anisotropic media. To lowest order, the anisotropic matter properties pertinent here are the electric and magnetic susceptibilities that were introduced in Chapter 5 to describe the local response of the material to electric and magnetic fields. The structure of these second rank symmetric tensors in the various crystal symmetries and their role in a number of static phenomena have already been discussed. The new aspects of crystal physics brought out in this chapter concern the effects of the anisotropy of (ϵ_{ij}) and (μ_{ij}) on the wave modes allowed to propagate in crystals.

We expect that in a homogeneous medium the basic wave modes are described by *plane waves* having the phase structure

$$\exp\ [i(\mathbf{k}\cdot\mathbf{r}-\omega t)] \qquad (9\text{-}1)$$

where $\mathbf{k}$ is the *wave vector* defining the spatial periodicity of the wave of frequency ω. Fronts of constant phase travel along the direction of $\mathbf{k}$ with a *phase velocity* c_n given by

$$c_n(\mathbf{k},\omega) = \omega/k = c/n \qquad (9\text{-}2)$$

where c is the velocity of light in free space and n is the traditional index of refraction. As indicated in Eq. (9-2), n is a function of both $\mathbf{k}$ and ω.

In isotropic materials n is independent of direction and the allowed modes in a given direction of travel are two linearly independent waves differing in their state of polarization but having a common velocity c_n. The two modes are degenerate in the sense that the two independent states of polarization can be chosen arbitrarily. The major consequence of crystal anisotropy is that this degeneracy of the modes is lifted. The index of refraction n is now a double-valued function of the direction of propagation, with one value for each of two modes of well-specified polarization.

In this chapter we develop the construction of the proper modes of electromagnetic waves in different crystal systems, and discuss their properties. Optical activity and other higher-order optical effects are taken up in Chapter 10.

9.1. GENERAL PROPERTIES OF NORMAL MODES

In nonconducting media, Maxwell's equations are

$$\nabla \times \mathbf{H} = \frac{\partial \mathbf{D}}{\partial t}, \quad \nabla \times \mathbf{E} = -\frac{\partial \mathbf{B}}{\partial t} \tag{9-3}$$

If we apply these equations to the plane waves of Eq. (9-1), and designate the *wave amplitudes* by the vectors **H**, **E**, **D**, and **B**, we obtain the set of relations

$$\mathbf{k} \times \mathbf{H} = -\omega \mathbf{D}, \quad \mathbf{k} \times \mathbf{E} = \omega \mathbf{B} \tag{9-4}$$

It follows from Eq. (9-4) that **D** and **B** are normal to the direction of propagation given by **k**. **D** is also normal to **H**, and **B** to **E**. Neither **E** nor **H** is generally transverse to **k**, and **B** is not necessarily normal to **D**. But since at optical frequencies the magnetic permeability is usually isotropic and equal to its value μ_0 in free space, in practice the relations between these vectors are somewhat simpler. With **B** and **H** having a common direction, a typical arrangement of the field vectors is shown in Fig. 9-1. The vectors **E**, **D**, and **k** are in the plane normal to **H**, but **E** is not a parallel to **D**.

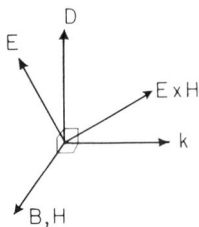

Fig. 9-1 Directions of the vectors of Eq. (9-4) if **B** and **H** are parallel.

The relation between **E** and **D** is specified by the dielectric tensor ϵ_{ij} in the equation

$$D_i = \sum_j \epsilon_{ij} E_j \qquad (9\text{-}5)$$

This equation is formally the same as Eq. (5-3) of Chapter 4. However, since the construction of waves based on the complex form of Eq. (9-1) leads to complex wave amplitudes **D** and **E**, the (ϵ_{ij}) of Eq. (9-5) must also be considered a complex quantity.

If the medium transmits waves without attenuation, the intrinsic symmetry of (ϵ_{ij}) is given by

$$\epsilon_{ij} = \epsilon_{ji}^* \qquad (9\text{-}6)$$

such that ϵ_{ij} is *Hermitian* if field energy is conserved. Invariance of this energy under time reversal requires the further symmetry

$$\epsilon_{ij}(-\mathbf{k}) = \epsilon_{ji}(\mathbf{k}) \qquad (9\text{-}7)$$

which becomes important in discussing spatial dispersion.

The direction of energy flow is defined by that of the *Poynting vector*, proportional to **E** × **H**. As indicated in Fig. 9-1, this direction differs from that of **k** if **E** and **D** are not parallel.

9.2. FRESNEL FORMULAS

The Fresnel formulas give the phase velocity c_n as a function of the direction of **k**, and establish the normal modes appropriate to **k**.

Without loss of generality we can choose as our frame of reference the principal axis system of (ϵ_{ij}). In this coordinate system the matter relations between the field vectors are

$$D_i = \epsilon_i E_i, \qquad B_i = \mu_0 H_i \qquad (9\text{-}8)$$

The allowed modes and propagation velocities are obtained by solving the set of equations (9-4) and (9-8) for the vector field amplitudes for a given direction of **k**. As in all problems of normal modes, we are led to a set of linear homogeneous equations that is soluble only if a secular equation is satisfied. This equation then specifies the phase velocities of Eq. (9-2) for the given direction of **k**.

Rather than carry out this standard procedure to find the secular equation, we employ an elegant shortcut permitted here by the particular form of Eq. (9-4). Eliminating **B** and **H** in Eq. (9-4), we obtain the expression

$$\mathbf{D} = \frac{k^2}{\mu_0 \epsilon_0 \omega^2} \left\{ \epsilon_0 \mathbf{E} - \frac{\epsilon_0}{k^2} (\mathbf{E} \cdot \mathbf{k})\mathbf{k} \right\} \qquad (9\text{-}9)$$

Introducing the principal axis velocities

$$c_i^2 = (\epsilon_0/\epsilon_i)c^2 \tag{9-10}$$

and using Eq. (9-2), we can cast Eq. (9-9) into the component form

$$D_i(c_n^2 - c_i^2) = -c^2 \frac{\epsilon_0(\mathbf{E} \cdot \mathbf{k})}{k^2} k_i \tag{9-11}$$

The orthogonality of **D** and **k**, applied to Eq. (9-11), leads to the secular equation

$$\sum_i \frac{k_i^2/k^2}{c_n^2 - c_i^2} = 0 \tag{9-12}$$

Equation (9-12) can be resolved to give an equation quadratic in c_n^2

$$c_n^4 - [(1-K_1^2)c_1^2 + (1-K_2^2)c_2^2 + (1-K_3^2)c_3^2]c_n^2 \tag{9-13}$$
$$+ [K_1^2 c_2^2 c_3^2 + K_2^2 c_3^2 c_1^2 + K_3^2 c_1^2 c_2^2] = 0$$

where we have introduced the direction cosines of **k**

$$K_i = k_i/k \tag{9-14}$$

As expected, for a given direction **K**, Eq. (9-13) has in general two distinct solutions that define the velocities of the two normal modes. The amplitudes of **D** of these normal modes follow directly from Eq. (9-11)

$$D_1 : D_2 : D_3 \tag{9-15}$$
$$= (c_n^2 - c_2^2)(c_n^2 - c_3^2)K_1 : (c_n^2 - c_3^2)(c_n^2 - c_1^2)K_2 : (c_n^2 - c_1^2)(c_n^2 - c_2^2)K_3$$

and the magnetic field corresponding to Eq. (9-15) is given by Eq. (9-4)

$$H_1 : H_2 : H_3 = -c_n(c_n^2 - c_1^2)(c_2^2 - c_3^2)K_2 K_3 :$$
$$-c_n(c_n^2 - c_2^2)(c_3^2 - c_1^2)K_3 K_1 : -c_n(c_n^2 - c_3^2)(c_1^2 - c_2^2)K_1 K_2 \tag{9-16}$$

The two remaining fields **E** and **B** are obtained by using Eq. (9-8).

The solution is completed by defining the *direction of energy* propagation $\hat{\mathbf{K}}$. Since this is the direction of the Poynting vector, substitution of the factors of Eqs. (9-15) and (9-16) for the field amplitudes gives the proportionalities

$\hat{K}_1 : \hat{K}_2 : \hat{K}_3$

$$= -K_1(c_n^2 - c_1^2)[(c_n^2 - c_3^2)^2(c_1^2 - c_2^2)c_2^2K_2^2 - (c_n^2 - c_2^2)^2(c_3^2 - c_1^2)c_3^2K_3^2] : -K_2(c_n^2 - c_2^2)[(c_n^2 - c_1^2)^2(c_2^2 - c_3^2)c_3^2K_3^2 - (c_n^2 - c_3^2)^2(c_1^2 - c_2^2)c_1^2K_1^2] : -K_3(c_n^2 - c_3^2)[(c_n^2 - c_2^2)^2(c_3^2 - c_1^2)c_1^2K_1^2 - (c_n^2 - c_1^2)^2(c_2^2 - c_3^2)c_2^2K_2^2] \quad (9\text{-}17)$$

Strictly speaking, the derivation of Eq. (9-13) is valid only if none of the factors in Eq. (9-11) vanish. However, since the general solutions of Eq. (9-13) must go over continuously into those applying in the special situations where some of the factors in Eq. (9-11) vanish, Eq. (9-13) defines a full set of solutions.

This is not so for Eqs. (9-15), (9-16), and (9-17), because when the proportionality contains common factors that vanish, the remaining factors are not necessarily defined. In that case, the normal mode amplitudes must be determined directly from the defining equations, such as Eq. (9-4). The detailed procedure for this is explained by the specific examples taken up in the next sections, where we treat separately the two major classes of optically anisotropic crystals characterized by the inequality of two or three of the principal dielectric constants.

9.3. UNIAXIAL CRYSTALS

In uniaxial crystals two of the three principal dielectric constants are equal. Hence the plane defined by two principal axes is dielectrically isotropic, and the unique principal direction is perpendicular to the isotropic plane. Uniaxial crystals act, to lowest order, like cylindrically symmetric media. The crystal symmetries leading to uniaxial dielectric behavior have been discussed in Section 5-2.

If we choose $\epsilon_1 = \epsilon_2$, and $\epsilon_3 \neq \epsilon_1$, the z direction defines the axis of symmetry. Because of this symmetry, the optical properties depend only on the angle of **K** with respect to the principal axis. It is, however, best to use a coordinate system independent of **K**.

With $\mathbf{K} = (K_1, K_2, K_3)$, and

$$c_1 = c_2 = c/(\epsilon_1)^{1/2}, \quad c_3 = c/(\epsilon_3)^{1/2}$$

Equation (9-13) factors to give the solutions

$$c_n^2 = c_1^2, \quad c_n^2 = c_3^2(K_1^2 + K_2^2) + c_1^2 K_3^2 \quad (9\text{-}18)$$

The first solution is isotropic. The second depends on the direction of propagation. Both solutions coalesce when **K** lies along the axis of symmetry, but differ for all other directions. As shown in Fig. 9-2, the anisotropic phase velocity in the xz plane may be larger or smaller than c_1, depending on the relative size of ϵ_1 and ϵ_3. The graphical solution for all directions of **K** is obtained by rotating the curves of Fig. 9-2 around the z axis. The nonspherical surfaces are surfaces of fourth order.

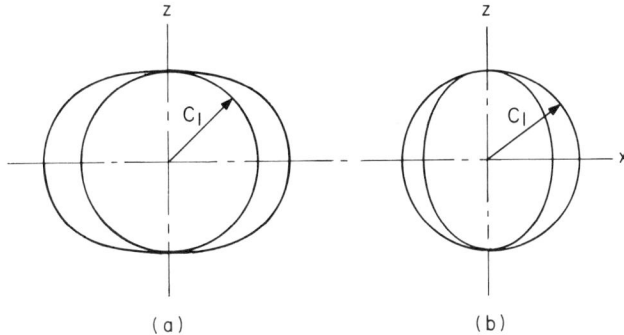

Fig. 9-2 Radial plot of c_n versus **K** for uniaxial crystals, according to Eq. (9-18). (a) $c_3 > c_1$; (b) $c_3 < c_1$.

The mode amplitudes belonging to the anisotropic phase velocity follow from Eqs. (9-15) and (9-16) by substituting the value of c_n^2 from Eq. (9-18), and equating c_1 and c_2:

$$D_1 : D_2 : D_3 = -K_1 K_3 : -K_2 K_3 : (K_1^2 + K_2^2) \tag{9-19}$$

$$H_1 : H_2 : H_3 = K_2 : -K_1 : 0$$

Hence, for the anisotropic mode, **D** lies in the plane defined by **K** and the z axis, and **H** lies in the xy plane. For given **K**, this mode has a well-defined plane polarization.

The amplitudes of the isotropic mode must be obtained by going back to Eq. (9-11). The condition $c_n^2 = c_1^2 = c_2^2$ leads to the two requirements

$$(\mathbf{E} \cdot \mathbf{K}) = 0, \quad D_3 = 0$$

These two requirements, together with Eq. (9-4), fully define the mode amplitudes

$$D_1 : D_2 : D_3 = -K_2 : K_1 : 0$$
$$H_1 : H_2 : H_3 = -K_1 K_3 : -K_2 K_3 : (K_1^2 + K_2^2)$$
(9-20)

As expected from the isotropy of c_n, the **D** of this mode lies in the isotropic xy plane, while **H** is normal to **D** and **K**. This mode is also plane polarized, and it is orthogonal to that of Eq. (9-19).

We conclude that for a given direction of propagation **K** the uniaxial crystal crystal supports two uniquely defined plane-polarized normal modes orthogonal to one another. Their directions of polarization are shown in Fig. 9-3.

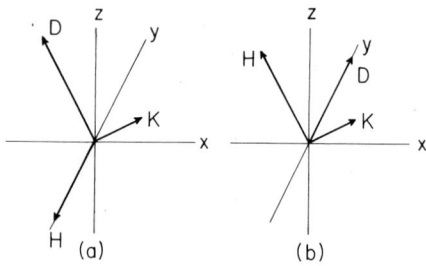

Fig. 9-3 Field directions of the two modes of propagation in a uniaxial crystal for (a) the anisotropic, and (b) the isotropic mode.

Each mode has its distinct group velocity and direction of energy propagation. For the isotropic mode, clearly $c_r = c_1$ and $\hat{\mathbf{K}} = \mathbf{K}$. For the anisotropic mode, Eq. (9-17) leads to the direction $\hat{\mathbf{K}}$.

$$\hat{\mathbf{K}} = (c_3^4 (K_1^2 + K_2^2) + c_1^4 K_3^2)^{-1/2} [c_3^2 K_1, c_3^2 K_2, c_1^2 K_3] \quad (9-21)$$

and the group velocity c_r

$$c_r = \left\{ \frac{c_3^4 (K_1^2 + K_2^2) + c_1^4 K_3^2}{c_3^2 (K_1^2 + K_2^2) + c_1^2 K_3^2} \right\}^{1/2} \quad (9-22)$$

According to Eq. (9-21), $\hat{\mathbf{K}}$ lies in the plane formed by $\mathbf{K}$ and the z axis. If $c_3 < c_1$, $\hat{\mathbf{K}}$ ic closer to the z axis than $\mathbf{K}$, but for $c_3 > c_1$ the order of these directions is reversed. It is seen from Fig. 9-2 that there is no simple relation between the phase velocity surface and the direction of $\hat{\mathbf{K}}$.

For propagation along the z axis, the two phase velocities are equal and the two modes become *degenerate*, so that all polarization states are transmitted without modification. Furthermore, it follows from Eqs. (9-21) and (9-22) that the flow of energy is also in the same direction, with $c_r = c_n$. The z axis defines the unique *optic axis* of uniaxial crystals. Higher-order optical effects may or may not lift the mode degeneracy along this axis. As an example of such perturbation, Problem 9-9 discusses the effect of a slight magnetic anisotropy on the normal modes.

9.4. BIAXIAL CRYSTALS

In biaxial crystals all three principal dielectric constants differ from each other. Hence, none of the three principal crystal axes can be singled out for characterizing optical behavior. In fact, the symmetries of optical properties in such crystals must be dependent on the specific numerical values of the dielectric constants.

The velocities of propagation and the amplitudes of the allowed modes in such a medium are fully described by the equations of Section 9-2. To gain some insight into the nature of the actual modes it is convenient to examine special directions of propagation. Let us restrict $\mathbf{K}$ to lie in one of the three principal planes. For these values of $\mathbf{K}$, Eq. (9-13) assumes the formal structure already encountered in the uniaxial case, and can be factored. Its pairs of solutions are

$$K_1 = 0, \quad c_n^2 = c_1^2, \quad c_n^2 = c_3^2 K_2^2 + c_2^2 K_3^2$$

$$K_2 = 0, \quad c_n^2 = c_2^2, \quad c_n^2 = c_3^2 K_1^2 + c_1^2 K_3^2 \quad (9\text{-}23)$$

$$K_3 = 0, \quad c_n^2 = c_3^2, \quad c_n^2 = c_2^2 K_1^2 + c_1^2 K_2^2$$

The solutions show the same structure and angular dependence found in Eq. (9-18) for the uniaxial case, and we expect that the corresponding normal modes follow the same construction already used in Section 9-3. The only difference is that since all three principal phase velocities are distinct, the points of degeneracy along the principal axis are removed.

The phase velocities of Eq. (9-23) are shown graphically in Fig. 9-4, where we have made the assumption

$$\epsilon_1 < \epsilon_2 < \epsilon_3, \quad c_1 > c_2 > c_3$$

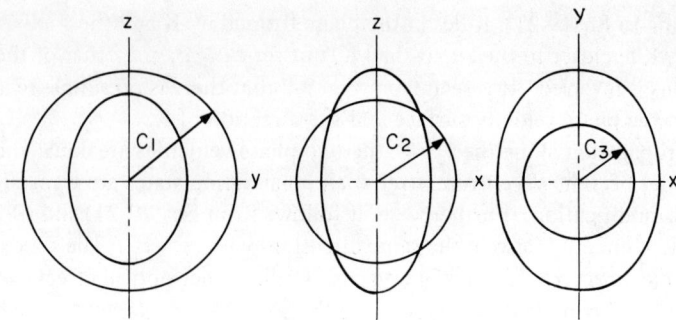

Fig. 9-4 Radial plot of c_n versus **K** for biaxial crystals, with **K** in one of the three principal planes. The plots follow from Eq. (9-23), with $c_1 > c_2 > c_3$.

The new feature of these curves is the appearance of the degeneracy of the two phase velocities in the xz plane. The common phase velocities $c_n = c_2$ occur along the directions

$$\mathbf{K} = \left[\pm \left(\frac{c_1^2 - c_2^2}{c_1^2 - c_3^2} \right)^{1/2}, \; 0, \; \pm \left(\frac{c_2^2 - c_3^2}{c_1^2 - c_3^2} \right)^{1/2} \right] \qquad (9\text{-}24)$$

These directions define the two symmetrically placed optic axes of the biaxial crystal. Just as in the uniaxial case, along these axes the two plane-polarized modes are degenerate and can be superimposed, with a common phase velocity, in any linear combination. In contrast to the uniaxial case, however, the energy flow of such a wave is not necessarily along the direction given by Eq. (9-24), but depends on the composition and polarization of the mode. For example, the mode deriving from the isotropic mode with $c_n = c_2$ sends energy along the optic axis, while the mode identified with the anisotropic surface of c_n transmits energy into the direction obtained by using in Eq. (9-17) the given **K** of the optic axes

$$\hat{\mathbf{K}} = \left\{ c_1^2(c_2^2 - c_3^2) + c_2^2 c_3^2 \right\}^{-1/2} \left[\pm c_3^2 \left(\frac{c_1^2 - c_2^2}{c_1^2 - c_3^2} \right)^{1/2}, \; 0, \right.$$

$$\left. \pm c_1^2 \left(\frac{c_2^2 - c_3^2}{c_1^2 - c_3^2} \right)^{1/2} \right] \qquad (9\text{-}25)$$

which is always between the z direction and an optic axis. As worked out in Problem 9-10, arbitrary combinations of the two modes send energy along directions lying on a circular cone between the two extremes defined by Eqs. (9-24) and (9-25). Thus, an unpolarized beam sent into a biaxial crystal along the optic axis will suffer *conical refraction*. For a quantitative theory of the intensity distribution of the light in the cone it is necessary to consider in detail the finite lateral dimensions of such an incoming beam.

It can be shown that the directions of degeneracy given by Eq. (9-24) are the only such directions. To visualize this it is useful to combine the solutions of the three principal planes shown in Fig. 9-4 into one plot exhibiting their interconnections. Figure 9-5 gives one octant of the surfaces of phase velocities and shows that the two sheets are connected at the four points of degeneracy in the xz plane. There is, therefore, a continuous transition between the velocities and the modes of the two sheets as the direction of $\mathbf{K}$ is changed.

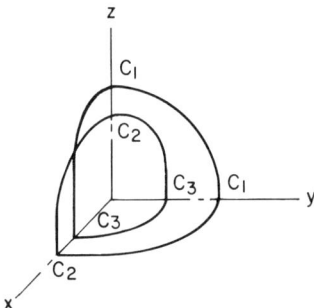

Fig. 9-5 Superposition of the cuts of Fig. 9-4 in three dimensions. The surface of phase velocities intersects itself to produce two sheets.

As indicated at the beginning of this section, the directions of the optic axes, while confined to the plane defined by the largest and smallest principal dielectric constants, are determined by the actual numerical values of ϵ_1, ϵ_2, and ϵ_3. Hence, we find that in biaxial crystals there is a dispersion of optic axes if the three dielectric constants show different dependences on ω.

9.5. COUPLING TO AN ISOTROPIC MEDIUM

In the preceding sections we established the electromagnetic wave modes allowed to propagate in an anisotropic medium. The *excitation* of these modes by an incident wave depends on the boundary conditions that all electromagnetic fields have to satisfy at the surface of the crystal. Maxwell's equations specify these conditions in terms of the normal or tangential components of the field vectors. Furthermore, consistent compliance with the boundary condition at different points of an interface requires the *continuity of phase* of the wave of Eq. (9-1). This continuity is expressed by conservation of the component of **k** in the interface, or correspondingly by

$$\left[\frac{1}{c}\mathbf{K}_\parallel\right]_{\text{outside}} = \left[\frac{1}{c_n(\mathbf{K})}\mathbf{K}_\parallel\right]_{\text{inside}} \qquad (9\text{-}26)$$

With c as the velocity of light in vacuum, the left-hand side of Eq. (9-26) specifies the direction and angle of incidence of the incoming wave. On the right side c_n differs from c and is a function of **K**. Therefore, Eq. (9-26) defines the allowed value of **K**, one for each of the two sheets of phase velocities, that couples to the external wave. Once the directions of wave propagation are established, the amplitudes of excitation of the modes follow from the continuity requirements of the incident, reflected, and transmitted electromagnetic fields. The continuity of the normal component of **D** is the most important condition.

In general, then, an incident wave excites two modes with different phase velocities and directions of propagation. The law of refraction in crystals, which defines the direction of the refracted beams, follows when the values of **K** allowed by Eq. (9-26) are related to the corresponding values of $\hat{\mathbf{K}}$. Hence there is double refraction for an unpolarized incident beam, with the direction of the anisotropic rays usually not lying in the plane of incidence. Some examples of the laws of refraction are handled in Problems 9-13 and 9-14. Obviously, the detailed solution of these laws in the general case for biaxial crystals is very involved. The principle of any solution, however, is fully contained in Eq. (9-26).

Equation (9-26) also applies, of course, to waves leaving the crystal. A single mode inside the crystal gives rise to a single refracted wave of specific polarization. If, however, the two internal modes resulting from the refraction of an incident wave still overlap spatially at the exit surface, they will recombine there to form waves in free space that usually have new states of polarization. The difference in phase velocities and propagation lengths of the two internal modes produces a relative change in their phases at the exit surface which enters into the continuity conditions for the field vectors. This feature of the propagation of light in crystals has led to numerous optical devices for analyzing or modifying the polarization state of a wave.

Problems

9-1. Show that the reciprocity relations of Eq. (9-7) imply that the basic optical phenomena do not distinguish between the presence or absence of the symmetry operation $\bar{1}$ in a crystal.

9-2. (a) Show that the secular equation Eq. (9-13) for the dispersion relation can be written for the index of refraction n of Eq. (9-2) the form

$$\epsilon_0^2 \left(\sum_i \epsilon_i K_i^2 \right) n^4 - \epsilon_0 \left(\sum_{i,j>i} \epsilon_i \epsilon_j (K_i^2 + K_j^2) \right) n^2 + \epsilon_1 \epsilon_2 \epsilon_3 = 0$$

(b) Show that if Eq. (9-8) is modified to have the form

$$B_1 = \mu_0 H_1, \qquad B_2 = \mu_0 H_2, \qquad B_3 = \mu_0 (1 + \delta) H_3$$

the secular equation becomes, to first order in δ

$$\epsilon_0^2 (1 - (K_1^2 + K_2^2)\delta) \left(\sum_i \epsilon_i K_i^2 \right) n^4$$
$$- \epsilon_0 \left[\sum_{i,j>i} \epsilon_i \epsilon_j (K_i^2 + K_j^2) - \epsilon_3 (\epsilon_1 K_1^2 + \epsilon_2 K_2^2)\delta \right] n^2$$
$$+ \epsilon_1 \epsilon_2 \epsilon_3 = 0$$

(c) Show that if for $\delta = 0$ in (a) the solutions are $n^2 = n_1^2, n_2^2$, then to first order in δ, the solutions in (b) are

$$n^2 = n_1^2 - \frac{n_1^2}{n_1^2 - n_2^2} \left[\frac{\epsilon_3}{\epsilon_0} \frac{\epsilon_1 K_1^2 + \epsilon_2 K_2^2}{\sum_i \epsilon_i K_i^2} - n_1^2 (K_1^2 + K_2^2) \right] \delta$$

$$n^2 = n_2^2 + \frac{n_2^2}{n_1^2 - n_2^2} \left[\frac{\epsilon_3}{\epsilon_0} \frac{\epsilon_1 K_1^2 + \epsilon_2 K_2^2}{\sum_i \epsilon_i K_i^2} - n_2^2 (K_1^2 + K_2^2) \right] \delta$$

9-3. Show, using Fig. 9-1, that the direction of energy flow $\hat{K}$ is in the plane of E, D and $K (=k/k)$. Furthermore, show that the velocity of energy flow is given by

$$c_r = \frac{c_n}{\mathbf{K} \cdot \hat{\mathbf{K}}}$$

and prove the inequality

$$c_n < c_r < c$$

9-4. Derive Eq. (9-17) by identifying the direction of $\hat{K}$ with that of $E \times H$.

9-5. (a) Show that, corresponding to the form of Eq. (9-4) given by

$$\mathbf{K} \times \mathbf{H} = -c_n \mathbf{D}, \qquad \mathbf{K} \times \mathbf{E} = c_n \mathbf{B}$$

we also have the equations

$$\hat{K} \times B = -\frac{1}{c_r} E, \quad \hat{K} \times D = \frac{1}{c_r} H$$

(b) Applying the procedure of Section 9-2 to these equations, show that corresponding to Eq. (9-11) we have

$$E_i \left(\frac{1}{c_r^2} - \frac{1}{c_i^2} \right) = -\frac{1}{\epsilon_0 c^2} (\hat{K} \cdot D) \hat{K}_i$$

and that the secular equation for the group velocity c_r is

$$\sum_i \frac{\hat{K}_i^2}{\left(1/c_r^2 - 1/c_i^2 \right)} = 0$$

9-6. Suppose that we are in an optically dispersive region characterized by

$$\epsilon_1 = \epsilon_2 > 0, \quad \epsilon_3 < 0.$$

(a) Determine the phase velocities of the normal modes in this medium, and plot the surfaces of phase velocities corresponding to Fig. 9-2.
(b) Show that the medium supports the isotropic mode with velocity c_1.
(c) Show that the anisotropic mode has an acceptable phase velocity c_n within a cone of directions around the z axis, but that on the basis of its group velocity and the direction of energy flow, this mode is not physical.

9-7. Show that the group velocity can be derived from the phase velocity c_n by

$$c_r = | \nabla_K c_n(K) |$$

and verify this formula by determining c_r of a uniaxial crystal, Eq. (9-22).

9-8. (a) Show that for a uniaxial crystal the secular equation for n of Problem 9-2 factors into the two equations

$$n^2 - (\epsilon_1/\epsilon_0) = 0, \quad [\epsilon_1(K_1^2 + K_2^2) + \epsilon_3 K_3^2] n^2 - (\epsilon_1 \epsilon_3/\epsilon_0) = 0$$

(b) Show that the surfaces defined by the *index of refraction vector* (nK_1, nK_2, nK_3) are ellipsoids of revolution, and compare their shapes to those of the construction of Fig. 9-3.

9-9. (a) Show that the perturbation introduced in Problem 9-2(b) does not lift the degeneracy of the mode along the optic axis of a uniaxial crystal.
(b) Show that to order δ the anisotropic mode is unaffected, but the isotropic mode has an index of refraction

$$n^2 = (\epsilon_1/\epsilon_0) [1 + (1 - K_3^2) \delta]$$

Crystal Optics   117

and determine the perturbed normal mode amplitudes replacing Eq. (9-20).

(c) Argue on physical grounds why the two modes should be affected differently by this perturbation.

9-10. Verify that the modes corresponding to the phase velocities in the principal planes of biaxial crystals, given by Eq. (9-23), agree with those derived for uniaxial crystals in Section 9-3.

9-11. Assume that the D fields deriving from the isotropic and anisotropic modes propagating along the optic axis of a biaxial crystal are in the proportion $\alpha : \beta$, with $\alpha^2 + \beta^2 = 1$.

(a) Show that the direction of energy flow of that wave is defined by

$$\hat{K}_0 = \left\{ 1 + \frac{(c_1^2 - c_2^2)(c_2^2 - c_3^2)}{c_2^4} \beta^2 \right\}^{-1/2}$$

$$\times \left[\left(\alpha^2 + \frac{c_3^2}{c_2^2}\beta^2\right)\left(\frac{c_1^2 - c_2^2}{c_1^2 - c_3^2}\right)^{1/2}, \; \alpha\beta \left(\frac{(c_1^2 - c_2^2)(c_2^2 - c_3^2)}{c_2^4}\right)^{1/2}, \right.$$

$$\left. \left(\alpha^2 + \frac{c_1^2}{c_2^2}\beta^2\right)\left(\frac{c_2^2 - c_3^2}{c_1^2 - c_3^2}\right)^{1/2} \right]$$

(b) Show that if to lowest order in the difference in velocities the common denominator of $\hat{K}_0$ can be set to unity, the projection of $\hat{K}_0$ in the plane perpendicular to the optic axis direction of Eq. (9-24) has the components

$$\hat{K}'_{0x} = -\left\{\frac{(c_1^2 - c_2^2)(c_2^2 - c_3^2)}{c_2^4}\right\}^{1/2} \beta^2, \quad \hat{K}'_{0y} = \left\{\frac{(c_1^2 - c_2^2)(c_2^2 - c_3^2)}{c_2^4}\right\}^{1/2} \alpha\beta$$

(c) Eliminate α and β to show that the equation relating $\hat{K}'_{0x}$ and $\hat{K}'_{0y}$ is a circle touching the direction of the optic axis.

9-12. Determine the direction of K for which the angle between K and $\hat{K}$ is maximized, if K lies in the principal plane containing the optic axis of a biaxial crystal. Find the value of the maximum angle, and show that it is larger than the maximum angle occurring in the other two principal planes.

9-13. Waves are incident on two surfaces S_1 and S_2 of a uniaxial crystal, as shown in Fig. 9-6.

(a) Show that K and $\hat{K}$ of all refracted waves lie in the xz plane.

(b) Show that the angles of refraction for wave normals and rays for the anisotropic mode are given by

$$S_1 : \tan\theta_n = \frac{(c_3/c)\sin\theta}{[1 - (c_1^2/c^2)\sin^2\theta]^{1/2}}, \quad \tan\theta_r = \frac{(c_1^2/c_3 c)\sin\theta}{[1 - (c_1^2/c^2)\sin^2\theta]^{1/2}}$$

$$S_2: \tan\theta_n = \frac{(c_1/c)\sin\theta}{[1-(c_3^2/c^2)\sin^2\theta]^{1/2}}, \qquad \tan\theta_r = \frac{(c_3^2/c_1 c)\sin\theta}{[1-(c_3^2/c^2)\sin^2\theta]^{1/2}}$$

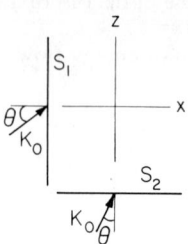

Fig. 9-6 Crystal surfaces S_1 and S_2 for Problem 9-13 of a uniaxial crystal with optic axis along z. The incident wave vector directions lie in the xz plane.

(c) Prove that, if $c_1 > c_3$, the anisotropic refracted ray is always closer to the z-direction than the refracted ray of the isotropic mode.

9-14. Assume that the incident wave on the surface S_1 of Problem 9-13 is not confined to the x-z plane, but has direction components

$$\mathbf{K}_{inc} = (\cos\theta, \sin\theta\sin\phi, \sin\theta\cos\phi)$$

Determine **K** and **K** of the refracted waves. Show that the refracted wave normals lie in the plane of incidence, but that the refracted anisotropic ray does not. Show that the angle at which this ray leaves the surface is given by

$$\tan\theta_r = \left\{ \frac{\left[\left(\frac{c_1}{c_3}\right)^2 \left(\frac{c_1}{c}\right)^2 \cos^2\phi + \left(\frac{c_3}{c}\right)^2 \sin^2\phi\right] \sin^2\theta}{1 - \left(\frac{c_1^2}{c^2}\cos^2\phi + \frac{c_3^2}{c^2}\sin^2\phi\right)\sin^2\theta} \right\}^{1/2}$$

9-15. Investigate the consequences on the propogating modes if the matter relations of Eq. (9-8) are modified by a magnetoelectric term of the form of Eq. (6-11).

Bibliography

J. F. Nye, *Physical Properties of Crystals,* Oxford Univ. Press, London and New York (1957), Chapter 13.

L. D. Landau and E. M. Lifshitz, *Electrodynamics of Continuous Media*, Addison-Wesley, Inc., Reading, Mass. (1960), Chapter 11.

M. Born and E. Wolf, *Principles of Optics*, MacMillan, New York (1959), Chapter 14.

W. P. Mason, *Crystal Physics of Interaction Processes*, Academic Press, New York (1966), Chapter 7.

CHAPTER 10

Second-Order Optical Effects

The two main features of electromagnetic waves propagating in anisotropic crystals derived in Chapter 9 are that for a given direction of propagation **K** there exist two orthogonal linearly polarized modes with distinct phase velocities and that for propagation along the optic axes these two modes become degenerate, so that in this case states of arbitrary polarization travel through the crystal without alteration. These results assume that the dielectric constant (ϵ_{ij}) of Eq. (9-5) is indeed a constant independent of other influences on the crystal, or that, at the most, it shows dispersion by varying with ω. To a high degree of approximation, this assumption is sufficient for describing the observed properties of light in crystals. However, (ϵ_{ij}) can be expected to show some sensitivity to other forces acting on the crystal, such as static electric or magnetic fields, which give the dielectric constant a new symmetry dependent on the direction of these forces.

In this chapter we explore some of the consequences of spatial dispersion of (ϵ_{ij}) and of the effect of static fields. In all such cases the basic modes of propagation developed in Chapter 9 are modified. Since the higher-order effects are small, any modifications can be treated in the spirit of perturbation theory, as for instance by expanding the new modes in terms of the original ones. Another typical feature of perturbation theory is the removal of degeneracies. If it applies here, the modes propagating along the optic axes may, in fact, be modified to first order.

10.1. OPTICAL ACTIVITY: SYMMETRY CONSIDERATIONS

If there is *spatial dispersion*, the dielectric constant is a function of wavelength as well as of frequency. Hence (ϵ_{ij}) is dependent on **k**. To first order such dependence can be expressed by an expansion

10.1 Second-Order Optical Effects

$$\epsilon_{ij}(\mathbf{k}) = \epsilon_{ij} + \epsilon_0 \sum_l g_{ijl} K_l \qquad (10\text{-}1)$$

where the proportionality of the right-hand term to the absolute value of k is absorbed in the coefficients (g_{ijl}). The dependence of the dielectric constant on the direction of propagation gives rise to *optical activity*. The nature of this effect is clarified by examining the properties of (g_{ijl}).

Applying the conditions of Eqs. (9-6) and (9-7) to Eq. (10-1), we find that (g_{ijl}) obeys the intrinsic symmetry

$$g_{ijl} = g_{jil}^* = -g_{jil} \qquad (10\text{-}2)$$

Consequently, it is a purely imaginary tensor antisymmetric in the first two indices. If we write

$$g_{ijl} = i\gamma_{ijl}, \qquad \gamma_{ijl} = -\gamma_{jil} \qquad (10\text{-}3)$$

the new coefficient (γ_{ijl}) has a structure similar to that of the Hall constant (R_{ijl}) discussed in Chapter 7. However, since **D**, **E**, and **k** are all polar, (γ_{ijl}) is a polar third rank tensor, while (R_{ijl}) is axial.

The presence of the factor i in Eq. (10-3) indicates that **D** and **E** are no longer in phase. The out-of-phase component introduces the possibility of elliptical or circular polarization in the normal modes. The linear dependence of (ϵ_{ij}) on the direction of propagation **K** ensures that the sense of rotation of such polarization relative to **K** remains unchanged when the propagation direction is reversed. Hence the sense of the "screw" along which **D** moves is built into the crystal. It results from the arrangement of atoms or bonds in the basis of the crystal cell.

Traditionally, optical activity is discussed in terms of a second rank tensor (g_{ml}) that incorporates the antisymmetry of the index pair (ij) of (γ_{ijl}) explicitly. The two descriptions are related by

$$\gamma_{ijl} = \sum_m \epsilon_{ijm} g_{ml} \qquad (10\text{-}4)$$

where (ϵ_{ijm}) is the antisymmetric triple product introduced in Eq. (3-3b). The relation between the single index m and the index pair (ij) follows the convention established in Eq. (7-43) in connection with the Hall effect.

Moreover, (g_{ml}) is an *axial* second rank tensor. Its symmetry under rotations and inversions is identical to that of the magnetoelectric tensor (λ_{ij}) of Section 6-4. Under an inversion symmetry $(\bar{1})$, (g_{ml}) goes into its negative. It follows that if $(\bar{1})$ is a symmetry operation of the crystal, no optical activity is allowed. If the inversion occurs in combination with some other symmetry elements, optical activity is not ruled out but may exist only in certain directions of propagation relative to these symmetry elements. In particular, whether optical activity is allowed along the optic axes depends on the specific location of these axes.

For example, consider the monoclinic crystal class (m), with the generating element m chosen such that the xz plane is the plane of reflection

$$m = \begin{pmatrix} 1 & 0 & 0 \\ 0 & -1 & 0 \\ 0 & 0 & 1 \end{pmatrix} \qquad (10\text{-}5)$$

This crystal is biaxial, and the x and z axes can always be chosen to be along the principal directions of (ϵ_{ij}). The tensor (g_{ml}), of course, is not constrained by this choice of axes. Applying the rules of Chapter 3 for reducing (g_{ml}) in order to satisfy the symmetry of Eq. (10-5), we obtain

$$(g_{ml}) = \begin{pmatrix} 0 & g_{12} & 0 \\ g_{21} & 0 & g_{23} \\ 0 & g_{32} & 0 \end{pmatrix} \qquad (10\text{-}6)$$

Working back through Eqs. (10-4) and (10-3) yields, for the form of the dielectric constant,

$$\begin{pmatrix} \epsilon_1 & i\epsilon_0 g_{32} K_2 & -i\epsilon_0 (g_{21} K_1 + g_{23} K_3) \\ -i\epsilon_0 g_{32} K_2 & \epsilon_2 & i\epsilon_0 g_{12} K_2 \\ i\epsilon_0 (g_{21} K_1 + g_{23} K_3) & -i\epsilon_0 g_{12} K_2 & \epsilon_3 \end{pmatrix} \qquad (10\text{-}7)$$

According to Eq. (10-7), a component of **D** generally couples to all three components of **E**. However, if **K** lies in the xz plane, only the field components in this plane interact, and the fields along y are completely unaffected by (g_{ml}). Hence, the propagating modes in the xz plane remain plane polarized. Therefore, if the optic axes lie in the xz plane, there will be no optical activity along their directions. On the other hand, if these axes lie in one of the other two coordinate planes, this restriction does not apply. The same conclusion could have been reached from general considerations of symmetry. If the direction of propagation lies in the plane of reflection, left- and right-handed elliptical polarizations must go into each other under the symmetry operation. This is possible only if the polarizations are linear.

As an example of optical activity in a uniaxial crystal, let us consider the crystal symmetry ($\overline{4}$). According to Appendix 3, the appropriate generating element is

10.1-10.2 Second-Order Optical Effects

$$(\bar{4}) = \begin{pmatrix} 0 & -1 & 0 \\ 1 & 0 & 0 \\ 0 & 0 & -1 \end{pmatrix} \qquad (10\text{-}8)$$

leading to the rotation tensor

$$\begin{pmatrix} g_{11} & g_{12} & 0 \\ g_{12} & -g_{11} & 0 \\ 0 & 0 & 0 \end{pmatrix} \qquad (10\text{-}9)$$

This tensor does not contribute to Eq. (10-1) when **K** lies along the optic axis, $\mathbf{K} = (0,0,1)$, but there are effects if K_1 and K_2 are nonvanishing.

In crystals not containing any inversion symmetry, (g_{ml}) has the form of a polar second rank tensor. If, as shown below, it is sufficient to consider symmetric tensors, (g_{ml}) in such crystals has the same scheme as (ϵ_{ij}). It also exists in isotropic media lacking inversion symmetry.

10.2. OPTICAL ACTIVITY: APPROXIMATE DISPERSION RELATIONS AND NORMAL MODES

Using the result of Problem 10-1(b), we have, for the relation between **D** and **E** in optically active media in the principal axis system of (ϵ_{ij}),

$$D_i = \epsilon_i E_i + i\epsilon_0 (\mathbf{E} \times \mathbf{G})_i \qquad (10\text{-}10)$$

and the relation between **D** and **E** from Maxwell's equations, in terms of the direction-dependent index of refraction n discussed in Problem 9-2, is

$$D_i = n^2 (\epsilon_0 E_i - \epsilon_0 (\mathbf{E} \cdot \mathbf{K}) K_i) \qquad (10\text{-}11)$$

These two equations are compatible only for special values of $n(\mathbf{K})$, and for definite proper modes. Since we expect that $\epsilon_0 G \ll \epsilon_i$, the new solutions for n and the mode amplitudes are obtainable by a perturbation of the known solutions for $\mathbf{G} = 0$.

By eliminating **D** between Eqs. (10-10) and (10-11) we obtain three equations for the electric field components.

$$E_i(\epsilon_i - n^2 \epsilon_0) + n^2 (\mathbf{E} \cdot \mathbf{K}) K_i - i\epsilon_0 (\mathbf{G} \times \mathbf{E})_i = 0, \quad i = 1,2,3 \qquad (10\text{-}12)$$

The secular determinant of these three homogeneous equations is

$$\epsilon_0^2\left(\sum_i \epsilon_i K_i^2\right) n^4 - \epsilon_0\left(\sum_{i,j>i} \epsilon_i\epsilon_j(K_i^2 + K_j^2) - \epsilon_0^2(\mathbf{K} \times \mathbf{G})^2\right) n^2$$
$$+ (\epsilon_1\epsilon_2\epsilon_3 - \epsilon_0^2 \sum_i \epsilon_i G_i^2) = 0 \qquad (10\text{-}13)$$

It specifies two characteristic solutions of n^2 appropriate to two independent proper modes.

If the characteristic solutions of n^2 in the absence of $\mathbf{G}$ are given by n_o^2 and n_e^2, Eq. (10-13) can be written to bring all terms dependent on $\mathbf{G}$ on the right-hand side

$$(n^2 - n_o^2)(n^2 - n_e^2) = \frac{\sum_i \epsilon_i G_i^2 - \epsilon_0 n^2(\mathbf{K} \times \mathbf{G})^2}{(\sum_i \epsilon_i K_i^2)} \qquad (10\text{-}14)$$

If on the right-hand side of Eq. (10-14) we make the consistent approximation

$$\epsilon_0 n^2 \simeq \epsilon_1 \simeq \epsilon_2 \simeq \epsilon_3$$

this side simplifies to become the scalar product $(\mathbf{K} \cdot \mathbf{G})^2$, which we designate by g^2. g is given by

$$g = (\mathbf{K} \cdot \mathbf{G}) = \sum_{m,l} g_{ml} K_m K_l \qquad (10\text{-}15)$$

In this approximation all optical activity depends only on the scalar function g. Because of the double summation in Eq. (10-15), any antisymmetric part of (g_{ml}) is suppressed, so that we have an argument for considering only symmetric tensors $g_{ml} = g_{lm}$.

Substituting Eq. (10-15) in Eq. (10-14), we obtain the modified solutions for the indices of refraction

$$\left.\begin{array}{c} n_o'^2 \\ n_e'^2 \end{array}\right\} = \frac{1}{2}\left((n_o^2 + n_e^2) \pm [(n_o^2 - n_e^2)^2 + 4g^2]^{1/2}\right) \qquad (10\text{-}16)$$

where the choice of sign is taken to give the proper limiting value of n^2 when g vanishes.

If $2g \ll |n_o^2 - n_e^2|$, Eq. (10-16) predicts that the indices of refraction are altered only to order g^2. Along an optic axis, however, where $n_o^2 = n_e^2$, the influence is of first order in g

$$\left.\begin{array}{c}n_o'^2\\ n_e'^2\end{array}\right\} = n_o^2 \pm g \qquad (10\text{-}17)$$

Hence, if g is nonvanishing for **K** along an optic axis, optical activity *removes the degeneracy* of the velocities in this direction. In all other directions of propagation the two already differing velocities are affected only to next higher order in g.

As a matter of fact, the validity of Eq. (10-17) requires further examination because some of the approximations used in deriving it are not obviously valid along the optic axes. Some aspects of this matter are taken up in Problems 10-5 to 10-7. The general conclusion is that in all cases where there is optical activity along the optic axes, Eq. (10-17) properly describes the difference in the indices of refraction to first order in **G** (assuming $\epsilon_1 = \epsilon_2 = \epsilon_3$ in the correction term), but that to second order, the average value of n^2, as well as the splitting of the indices of refraction, often shows additional terms. Furthermore, to second order in **G** the degeneracy in n^2 along the optic axes may be removed, although optical activity along these axes is forbidden by symmetry.

To interpret these results further we construct the modified normal modes of propagation. Using the approximation that led to Eq. (10-16), we can combine Eqs. (10-10) and (10-11) to obtain the composite equation

$$D_i\left(\frac{1}{n^2} - \frac{\epsilon_o}{\epsilon_i}\right) + \epsilon_o(\mathbf{E}\cdot\mathbf{K})K_i = -\frac{i}{n_{av}^4}(\mathbf{D}\times\mathbf{G})_i \qquad (10\text{-}18)$$

where n_{av} is an average index of refraction. We expand the fields in terms of the unperturbed solutions $(\mathbf{D}^o, \mathbf{E}^o)$ and $(\mathbf{D}^e, \mathbf{E}^e)$ corresponding to n_o^2 and n_e^2. The mode referring to $n_o'^2$ is anticipated to be of the form

$$\mathbf{D} = \mathbf{D}^o - i\alpha\mathbf{D}^e, \qquad \mathbf{E} = \mathbf{E}^o - i\alpha\mathbf{E}^e \qquad (10\text{-}19)$$

where α is a coefficient to be determined. The amplitudes of the unperturbed D fields can be chosen to be equal and at right angles

$$\mathbf{D}^o = \mathbf{D}^e \times \mathbf{K}$$

and each unperturbed solution obeys Eq. (10-18) if the right-hand side is set equal to zero and the proper value of n^2 is inserted.

Substituting Eq. (10-19) in Eq. (10-18), we find that the imaginary part of this equation is satisfied if

$$\alpha D_i^e\left(\frac{1}{n_o'^2} - \frac{1}{n_e^2}\right) = -\frac{1}{n_{av}^4}(\mathbf{D}^o\times\mathbf{G})_i \qquad (10\text{-}20)$$

and by writing Eq. (10-20) in terms of only one field, and using the orthogonality of **K** and **D**, we find that Eq. (10-20) leads to the value for α

$$\alpha = \frac{g}{n_e^2 - n_o'^2} \tag{10-21}$$

Equation (10-21) implies that in general the mixing of the modes is of order g. Along the optic axes, however, where $n_e^2 - n_o'^2 = -g$, $\alpha = -1$. It can be shown that this conclusion is consistent within all the approximations of this section. Consequently, because of the factor i of Eq. (10-19), the normal mode along an optic axis is *circularly polarized*. Away from an optic axis, the normal modes show elliptical polarization of order g.

The normal mode orthogonal to Eq. (10-19) is obtained by the replacements o → e, g → $-g$ in the foregoing derivation. These substitutions lead to the same value of α but because of the interchange of the two basic modes, the second mode represents elliptical or circular polarization in the opposite sense.

We conclude that within the approximations employed here, optical activity removes the degeneracy of the modes and of the propagation velocities along the optic axes. The modified modes along these directions are circularly polarized, whereas along other directions they form elliptically polarized waves.

As shown in Problems 10-9, 10-10, and 10-11, an exact solution of the normal modes along the optic axes indicates that to higher order in **G** these modes are not quite circularly polarized. In addition, in cases where the degeneracy in n^2 is removed without leading to optical activity, the perturbed modes remain linearly polarized. The extent to which these conclusions of a more exact treatment have significance depends, of course, on both the magnitude of the additional effects and the existence of competing higher-order terms.

10.3. EFFECTS OF STATIC MAGNETIC OR ELECTRIC FIELDS

The effect of a static magnetic field $\mathbf{B}^0$ on optical properties is a perturbation that can be described by the same formulation developed in Section 10-2 for optical activity. To first order in $\mathbf{B}^0$, the dielectric constant has the expansion

$$\epsilon_{ij}(\mathbf{B}^0) = \epsilon_{ij} + \epsilon_0 \sum_k r_{ijk} B_k^0 \tag{10-22}$$

The intrinsic symmetry of (r_{ijk}) follows from Eq. (9-6) and from the time-invariance conditions involving magnetic fields

$$\epsilon_{ij}(-\mathbf{B}^0) = \epsilon_{ji}(\mathbf{B}^0) \tag{10-23}$$

10.3 Second-Order Optical Effects

Thus, (r_{ijk}) is purely imaginary and antisymmetric in the first two indices. Consequently, we can define the auxiliary real coefficients (r_{ij}) and R_i by the equations

$$r_{ijk} = i \sum_l \epsilon_{ijl} r_{lk} \qquad (10\text{-}24)$$

$$R_i = \sum_k r_{ik} B_k^0 \qquad (10\text{-}25)$$

and the constitutive equation between **D** and **E** in the principal axis system of (ϵ_{ij}) has the form

$$D_i = \epsilon_i E_i + i\epsilon_0 (\mathbf{E} \times \mathbf{R})_i \qquad (10\text{-}26)$$

This form is fully analogous to Eq. (10-10), so that all the results of Section 10-2 can be taken over formally to describe the changes in the propagation properties and in the normal modes of electromagnetic waves in the presence of $\mathbf{B}^0$.

However, these generalized *Faraday effects* differ from optical activity in two important respects. First, the reversal of the direction of propagation $\mathbf{K} \to -\mathbf{K}$ does not change the sign of **R**. Hence the sense of rotation of elliptical or circular polarization relative to **K** reverses with the direction of propagation. Second, because **B** is an axial vector, (r_{ijk}) is an axial third rank tensor and therefore (r_{ij}) is a *polar second rank tensor*. Consequently, the restrictions of crystal symmetry on the Faraday effect differ from those on optical activity whenever inversion plays a role. In fact, the symmetry of (r_{ijk}) is formally identical with that of the Hall coefficient (R_{ijk}) of Chapter 7, and all the results concerning symmetry obtained there can be directly applied to the Faraday effect.

In summary, the presence of $\mathbf{B}^0$ generally leads to elliptically polarized modes of propagation and it removes the degeneracy of these modes along the optic axes. To first order in **R**, the modes along the optic axes are circularly polarized. Finally, the Faraday effect does not distinguish between crystal symmetries involving rotations or rotation–inversions.

In the presence of a static electric field $\mathbf{E}^0$, the expansion analogous to Eq. (10-22) is

$$\epsilon_{ij}(\mathbf{E}^0) = \epsilon_{ij} + \sum_k \beta_{ijk} E_k^0 \qquad (10\text{-}27)$$

and while the intrinsic symmetry requirement of Eq. (9-6) still applies, time reversal invariance now leads to the condition

$$\epsilon_{ij}(\mathbf{E}^0) = \epsilon_{ji}(\mathbf{E}^0) \tag{10-28}$$

Hence (β_{ijk}) is a real tensor symmetric in the first index pair. Consequently, Eq. (10-27) primarily describes a change in the value of the elements of the dielectric constant (ϵ_{ij}). The effect of $\mathbf{E}^0$, then, is to alter the magnitude and perhaps the symmetry of the dielectric tensor, thus changing the speeds of propagation of the allowed modes, and their degeneracies, without affecting the basic mode structure of two orthogonal linearly polarized waves. (β_{ijk}) is a polar third rank tensor symmetric in the first index pair. The detailed structure as restricted by crystal symmetry is fully analogous to that of the more familiar piezoelectric tensor taken up in Chapter 12.

10.4. FREE ENERGY FORMULATION OF OPTICAL EFFECTS

The higher-order effects treated in this chapter are so far based on a purely formal expansion of the dielectric constant in terms of various parameters. This formulation arbitrarily assigns all interactions with the medium to the electric fields, and ignores any magnetic couplings, even though in the electromagnetic wave both fields are on the same footing. A more symmetric approach is desirable because it would allow greater flexibility in the formulation of higher-order effects, especially those including more than one field, and because it should lead to a more systematic treatment of the intrinsic symmetry accompanying such effects.

The free energy formulation developed by Pershan contains all of these features. Let us assume that the material contains electric and magnetic dipole densities $\mathbf{P}$ and $\mathbf{M}$ at the frequency ω. Then with $\mathbf{B} = \mu_0 \mathbf{H} + \mathbf{M}$ and $\mathbf{D} = \epsilon_0 \mathbf{E} + \mathbf{P}$, the energy density U of the electromagnetic field due to the material obeys the equation

$$\frac{\partial U}{\partial t} = \mathbf{H} \cdot \frac{\partial \mathbf{M}}{\partial t} + \mathbf{E} \cdot \frac{\partial \mathbf{P}}{\partial t} \tag{10-29}$$

and represents the work done by the external fields on the system. For a periodic disturbance in the steady state the time average of Eq. (10-29) vanishes. However, there is an energy density associated with the establishment of the fields in the crystal, and if that energy depends only on the final state of the system, we must be able to write a perfect differential

$$dU = \mathrm{Re}(\mathbf{H}^* \cdot d\mathbf{M} + \mathbf{E}^* \cdot d\mathbf{P}) \tag{10-30}$$

where the symbols on the right-hand side now refer to the complex field amplitubes of the fields varying like $\exp(-i\omega t)$. Hence $U(\mathbf{M},\mathbf{P})$ is a function of state.

The corresponding free energy F in terms of the variables $\mathbf{H}^*$ and $\mathbf{E}^*$ is given by

$$F(\mathbf{H}^*,\mathbf{E}^*) = U - \mathrm{Re}(\mathbf{H}^* \cdot \mathbf{M} + \mathbf{E}^* \cdot \mathbf{P}) \qquad (10\text{-}31)$$

and the dependent variables $\mathbf{M}$ and $\mathbf{P}$ are derived from F by

$$M_i = -\frac{\partial F}{\partial H_i^*}, \qquad P_i = -\frac{\partial F}{\partial E_i^*} \qquad (10\text{-}32)$$

Thus, given the form of F, we obtain $\mathbf{M}$ and $\mathbf{P}$ from Eq. (10-32) and can then construct the traditional fields for solving Maxwell's equations, Eq. (9-3). All optical effects follow from different terms in F. As an example, let us consider a free energy density of the form

$$F = -\frac{\epsilon_0}{2} \sum_{i,j} (\alpha_{ij} E_i^* E_j + \alpha_{ij}^* E_i E_j^*) \qquad (10\text{-}33)$$

According to Eq. (10-32), the polarization density is given by

$$P_i = \frac{\epsilon_0}{2} \sum_j (\alpha_{ij} + \alpha_{ji}^*) E_j \qquad (10\text{-}34)$$

Since Eq. (10-33) contains only nine products $E_i E_j^*$, (α_{ij}) can have only nine independent components and must be Hermitian. Furthermore, invariance of F under time reversal $\mathcal{R}$ requires that (α_{ij}) be real. Thus we are led to the conventional real and symmetric polarizability tensor.

Similarly, if we assume a free energy term given by

$$F = -\frac{1}{2} \sum_{i,j} (\gamma_{ij} E_i^* H_j + \gamma_{ij}^* E_i H_j^*) \qquad (10\text{-}35)$$

we obtain a cross coupling between electric and magnetic effects

$$M_i = \sum_j \gamma_{ji}^* E_j, \qquad P_i = \sum \gamma_{ij} H_j \qquad (10\text{-}36)$$

where (γ_{ij}) is a general second rank tensor that is purely imaginary because the invariance of F under $\mathcal{R}$, together with the relations $\mathcal{R}\mathbf{E} = \mathbf{E}^*$, $\mathcal{R}\mathbf{H} = -\mathbf{H}^*$, requires $\gamma_{ij} = -\gamma_{ij}^*$.

Equation (10-36) describes an out-of-phase response of the material to applied electromagnetic fields, and thus corresponds to optical activity. It is not obvious, however, to what extent the details of this formulation agree with those of Eq. (10-1). This question is explored in Problem 10-14. Note, however, that here the origin of optical activity resides fundamentally in a coupling of electric and magnetic fields by the medium. Spatial dispersion is a result of that coupling rather than a position-dependent response to the electric field alone.

The free energy expressions constructed so far deal only with fields at one frequency, ω. However, the formalism can easily be extended to include more than one frequency. It will then describe the interactions between fields of different frequencies that form the basis of nonlinear optics. The various frequencies entering in such expressions are not independent. If three frequencies are involved, for example, we expect that there exists the conservation law $\omega_3 = \omega_1 + \omega_2$. In particular, if $\omega_1 = 0$, we obtain as a limiting case of nonlinear phenomena the interaction of an electromagnetic with a static field. Thus the effects discussed in Section 10-3 are part of the nonlinear response of crystals to electric and magnetic fields.

Problems

10-1. (a) Show that the dielectric constant of Eq. (10-1) can be written

$$\epsilon_{ij}(\mathbf{k}) = \epsilon_{ij} + i\epsilon_0 \sum_m \epsilon_{ijm} G_m$$

where $\mathbf{G}$ is a vector with components

$$G_m = \sum_l g_{ml} K_l$$

(b) Show that the relation between $\mathbf{D}$ and $\mathbf{E}$ takes the form

$$D_i = \sum_j \epsilon_{ij} E_j + i\epsilon_0 (\mathbf{E} \times \mathbf{G})_i$$

(c) Verify that the introduction of $\mathbf{G}$ does not affect the energy density of the electromagnetic field.

10-2. Discuss the possibility of elliptical polarization for a wave propagating in the plane of reflection, treated in Section 10-1, by using the vector $\mathbf{G}$ and the formula of Problem 10-1 relating $\mathbf{D}$ and $\mathbf{E}$.

10-3. (a) Determine the optical activity tensors g_{ml} and γ_{ijl} in the systems $(\bar{4}2m)$ and (mm).
(b) Establish whether there can be optical activity for propagation along the optic axes.

10-4. Verify Eq. (10-13).

10-5. Assume that light propagates along the optic axis of a uniaxial crystal: $\mathbf{K} = (0,0,1)$.

Second-Order Optical Effects

(a) Solve Eq. (10-13) exactly for this direction and show that the two indices of refraction are

$$n^2 = \frac{\epsilon_1}{\epsilon_0} - \frac{1}{2}\frac{\epsilon_0}{\epsilon_3}(G_1^2 + G_2^2) \pm \frac{1}{2}\frac{\epsilon_0}{\epsilon_3}\left[(G_1^2 + G_2^2)^2 + 4\frac{\epsilon_3^2}{\epsilon_0^2}G_3^2\right]^{1/2}$$

(b) Examine the structure of the vector **G** in all uniaxial crystals and show that the rigorous solutions are

$$n^2 = (\epsilon_1/\epsilon_0) \pm g_{33}$$

in agreement with Eq. (10-17).

10-6. Show that for light propagating along the optic axis of a biaxial crystal ($\epsilon_1 < \epsilon_2 < \epsilon_3$), whose directions are given by Eq. (9-24), the exact solution of Eq. (10-13) is

$$n^2 = \frac{\epsilon_2}{\epsilon_0} - \frac{1}{2}\frac{\epsilon_0 \epsilon_2}{\epsilon_1 \epsilon_3}[(G_2^2 + (K_3 G_1 - K_1 G_3)^2]$$

$$\pm \frac{1}{2}\frac{\epsilon_0 \epsilon_2}{\epsilon_1 \epsilon_3}\left\{\frac{4}{\epsilon_0^2}(\epsilon_1 K_1 G_1 + \epsilon_3 K_3 G_3)^2 + [G_2^2 + (K_3 G_1 - K_1 G_3)^2]^2\right\}^{1/2}$$

and establish that the values of **G** along the optic axis for a crystal of symmetry (222) lead to both a splitting and a shift of the average value of the indices of refraction.

10-7. Show that even though a crystal of symmetry (m) has no optical activity along the optic axes if these lie in the plane of reflection, the degeneracy of the indices of refraction for these propagation directions is removed. Use the results of Problem 10-6, and the form of (g_{ml}) of Eq. (10-6).

10-8. (a) Verify Eq. (10-21) in detail.
(b) Show that the real part of Eq. (10-18) corresponding to Eq. (10-20) leads to the condition

$$n_o'^2 - n_o^2 = -\alpha g$$

which confirms the consistency of the dispersion relation of Eq. (10-14).

10-9. (a) Show that if light propagates along the optic axis $\mathbf{K} = (K_1, 0, K_3)$, the ratio of the two components of **D** normal to **K**, $D_t = (K_3 D_1 - K_1 D_3)$ and D_2, is given exactly by

$$\frac{D_t}{D_2} = i\frac{-[(\epsilon_1/\epsilon_0)K_1 G_1 + (\epsilon_3/\epsilon_0)K_3 G_3] + iG_2(K_3 G_1 - K_1 G_3)}{\frac{1}{2}[-G_2^2 + (K_3 G_1 - K_1 G_3)^2] \pm \left\{\frac{1}{4}[G_2^2 + (K_3 G_1 - K_1 G_3)^2]^2 + [(\epsilon_1/\epsilon_0)K_1 G_1 + (\epsilon_3/\epsilon_0)K_3 G_3]^2\right\}^{1/2}}$$

(b) Show that these ratios correspond to two elliptically polarized modes having the same axis ratio but perpendicular to each other and rotating in opposite directions.
(c) Show that circular polarization is obtained only if $G_2 = 0$ and $(K_3 G_1 - K_1 G_3) = 0$.
(d) Verify that these two conditions are fulfilled identically in all uniaxial crystals, but that in biaxial crystals at best one or the other of these conditions holds. What does this imply about the exact normal modes along the *unperturbed* optic axes of a biaxial crystal?

10-10. Show that for the symmetry of Problem 10-7 the ratio of amplitudes of Problem 10-9 becomes indeterminate, and, by going back to a solution of normal modes based directly on Eq. (10-18), prove that the normal modes remain plane polarized.

10-11. According to Problem 10-9, the normal modes along the optic axes of biaxial crystals are not strictly circularly polarized.
(a) Show that a mode $\mathbf{D} = (K_3', i, -K_1')$ propagates along new directions $\mathbf{K}' = (K_1', 0, K_3')$ that satisfy the equation

$$K_1'^2 \left(\frac{\epsilon_1 \epsilon_2}{\epsilon_0^2} - G_3^2 \right) + 2 K_1' K_3' G_1 G_3 + K_3'^2 \left(\frac{\epsilon_2 \epsilon_3}{\epsilon_0^2} - G_1^2 \right)$$

$$- \left(\frac{\epsilon_1 \epsilon_3}{\epsilon_0^2} - G_2^2 \right) - 2 i G_2 (K_3' G_1 - K_1' G_3) = 0$$

(b) Show that these directions exist if $G_2 = 0$ and if

$$(\epsilon_1 \epsilon_3 / \epsilon_0^2)(\epsilon_3 - \epsilon_2)(\epsilon_2 - \epsilon_1) + G_3^2 \epsilon_3 (\epsilon_2 - \epsilon_1) - G_1^2 \epsilon_1 (\epsilon_3 - \epsilon_2) > 0$$

(c) Show that to lowest order in the components of $\mathbf{G}$, these directions are then given by

$$K_1'^2 = 1 - K_3'^2 = \frac{\epsilon_3(\epsilon_2 - \epsilon_1)}{\epsilon_2(\epsilon_3 - \epsilon_1)} \left\{ 1 - \epsilon_0^2 \frac{G_3^2 - G_1^2}{\epsilon_2(\epsilon_3 - \epsilon_1)} - \epsilon_0^2 \frac{G_1^2}{\epsilon_3(\epsilon_2 - \epsilon_1)} \right.$$

$$\left. \pm \epsilon_0^2 \frac{2 G_1 G_3}{\epsilon_2(\epsilon_3 - \epsilon_1)} \left[\frac{\epsilon_1(\epsilon_3 - \epsilon_2)}{\epsilon_3(\epsilon_2 - \epsilon_1)} \right]^{1/2} \right\}$$

where the plus sign goes with one optic axis, and the minus sign with the other. This indicates that the "experimental optic axes" are no longer symmetrical with respect to the z axis.
(d) Discuss what happens when $G_2 \neq 0$.

10-12. Derive the Faraday effect tensors in the crystal symmetries (m) and $(\overline{4})$ and compare the results to Eqs. (10-6) and (10-9).

10-13. Show that an isotropic medium can support a Faraday effect, but that a linear electro-optic effect is forbidden.

10-14. In order to connect the formulation of optical activity of Eq. (10-36) with that of Eq. (10-1) we must express all responses of the crystal in terms of electric fields, and use the vacuum values $\mathbf{B} = \mu_0 \mathbf{H}$ for the magnetic fields.

(a) Show that, using Eq. (9-4) and the general relations $\mathbf{B} = \mu_0 \mathbf{H} + \mathbf{M}$, $\mathbf{D} = \epsilon_0 \mathbf{E} + \mathbf{P}$, the equivalent $\mathbf{D}$ field is given by

$$\mathbf{D} = (\epsilon_0 \mathbf{E} + \mathbf{P} - \frac{\mathbf{k} \times \mathbf{M}}{\omega \mu_0})$$

(b) Show, using Eq. (10-36) and Eq. (9-2) with $\mathbf{K} = \mathbf{k}/k$, that

$$D_i = \epsilon_0 E_i + \epsilon_0 nc \sum_{j,k,l} [\gamma_{ij} \epsilon_{jkl} K_k E_l + \epsilon_{ijk} K_j \gamma_{lk} E_l]$$

(c) Compare with Eq. (10-1) and establish the relation

$$g_{ijl} = nc \sum_k (\gamma_{ik} \epsilon_{klj} + \epsilon_{ilk} \gamma_{jk})$$

and show the explicit one-to-one correspondence between the terms on both sides of this equation.

Bibliography

J. F. Nye, *Physical Properties of Crystals*, Oxford Univ. Press, London and New York (1957), Chapter 14.

L. D. Landau and E. M. Lifshitz, *Electrodynamics of Continuous Media*, Addison-Wesley, Inc., Reading, Mass. (1960), Chapter 11.

G. Szivessy, *Handbuch der Physik*, Vol. XX, Springer, Berlin (1928), Chapter 11.

P. S. Pershan, "Nonlinear Optical Properties of Solids: Energy Considerations," *Phys. Rev.* **130**, 919 (1963).

N. Bloembergen, *Nonlinear Optics*, W. A. Benjamin, Inc., New York (1965).

CHAPTER 11

Elasticity

When a crystal is subjected to mechanical forces acting on its free surfaces it changes its shape as well as its dimensions. As a response to the applied forces, the initial deformation is expected to be *elastic*, such that when the forces are removed the crystal recovers its original form. Furthermore, the symmetry of the deformation must reflect the symmetry of the undeformed crystal. In this chapter we develop the phenomenological theory of the elastic behavior of crystals in the limit of *infinitesimal homogeneous* deformations. In such deformations parallel lengths remain parallel and are changed in the same ratio, and equally positioned angles are changed by the same amount independent of their location relative to the origin of coordinates. Under static conditions the local distortion is the same throughout the crystal, because the forces acting on the free surfaces are transmitted by reaction forces to the surfaces of any elemental volume. If the forces are time dependent, the response of the crystal propagates these reactions with a finite velocity, thus leading to elastic waves. Both static and dynamic aspects of elasticity are taken up in this chapter.

11.1. THE ELASTIC PARAMETERS

Let us formulate the mathematical description of a homogeneous deformation. If the point (x,y,z) goes over into the point $(x + u_1, y + u_2, z + u_3)$, then we expect that u_i is a linear function of (x,y,z). The changes of scale along the coordinate axes are given by

$$\frac{\partial u_i}{\partial x_i} = e_{ii} \qquad (11\text{-}1)$$

and the change in angle between two orthogonal lines originally parallel to the axes defined by i and j is

$$\frac{\partial u_i}{\partial x_j} + \frac{\partial u_j}{\partial x_i} = e_{ij} \tag{11-2}$$

Since a rigid rotation of this pair of lines is described by the condition

$$\frac{\partial u_i}{\partial x_j} = -\frac{\partial u_j}{\partial x_i}$$

we can eliminate any pure rotation in the description of the deformation by imposing the additional requirement

$$\frac{\partial u_i}{\partial x_j} = \frac{\partial u_j}{\partial x_i} = \frac{1}{2} e_{ij} = \frac{1}{2} e_{ji} \tag{11-3}$$

Hence there are six parameters e_{ij}, the dimensionless components of *strain*, that fully define the homogeneous deformation

$$u_i = \sum_j \frac{1}{2}(1 + \delta_{ij}) e_{ij} x_j \tag{11-4}$$

The introduction of the factor $\frac{1}{2}$ in Eq. (11-3), traditional in many treatments of elasticity, leads to some minor complications in the manipulation of the strains. However, other complications of similar nature arise if these factors are omitted. In any case, these factors of $\frac{1}{2}$ call attention to the fact that in problems of elasticity different conventions are in use. A typical, though greatly exaggerated, deformation described by Eq. (11-4) is shown in Fig. 11-1.

Such deforming strains are caused by homogeneous forces acting on any element of area within the body. As shown in Fig. 11-2, the force acting on one such area element has a normal and two tangential components. The forces per unit area, the *stresses*, acting on three independent areas have altogether nine components. If the stresses are labeled by the two-index symbol X_{ij}, the first index gives the direction of the force, and the second indicates the normal of the surface on which the force acts. Among the nine stresses there exist the three relations

$$X_{ij} = X_{ji} \tag{11-5}$$

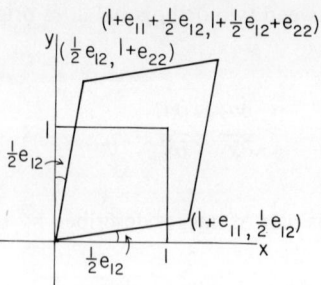

Fig. 11-1 Two-dimensional strain deforming a unit square according to Eq. (11-4)

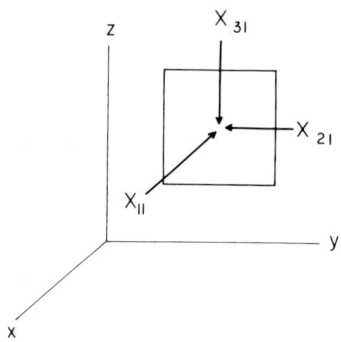

Fig. 11-2 Stresses on unit surface in the yz plane.

These equalities assure that there is no net torque of the shear stresses on any volume element. The normal or pressure stresses on this volume are in automatic balance in pairs acting on its opposite faces.

The index symmetry expressed by Eq. (11-3) and (11-5) allows us to introduce a single index, running from 1 to 6, to label the six independent strains and stresses e_i and X_i. The correspondence between single and double indices is that already introduced in Eq. (4-12) when labeling components of linear spaces of higher dimensions. $i = 1,2,3$ refers to normal strains or stresses, while $i = 4,5,6$ labels the shear components. In this shorthand notation, *Hooke's law* expresses the linear response of the medium in the familiar form

$$e_i = \sum_j s_{ij} X_j, \quad X_i = \sum_j c_{ij} e_j \tag{11-6}$$

The coefficients (s_{ij}) are the *elastic* or *compliance constants*, while the coefficients (c_{ij}) are the *elastic moduli* or *stiffness constants*. The two sets of coefficients are inverse to each other:

$$\sum_j s_{ij} c_{jk} = \sum_j c_{ij} s_{jk} = \delta_{ik} \tag{11-7}$$

Of the 36 entries in the array (c_{ij}), only 21 are independent. Since the deformations under study are infinitesimal and reversible, they are related to an elastic energy density $U(e_i)$, such that the stresses are given by

$$X_i = \partial U / \partial e_i \tag{11-8}$$

Equality of the second cross-derivatives of U implies the symmetry

$$c_{ij} = c_{ji}, \quad s_{ij} = s_{ji} \tag{11-9}$$

leading to 6 diagonal and $\frac{1}{2}(36-6) = 15$ independent off-diagonal elements, for a total of 21. The energy density U is then explicitly

$$U = \frac{1}{2} \sum_{i,j} c_{ij} e_i e_j = \frac{1}{2} \sum_{i,j} s_{ij} X_i X_j \tag{11-10}$$

11.2. TRANSFORMATION PROPERTIES AND SYMMETRY CONSIDERATIONS

The transformation properties of the elastic parameters are related directly to those of the coordinates. To establish this relation it is appropriate to employ the full index notation for stresses, strains, and elastic parameters. Hooke's law of Eq. (11-6) then takes the form

$$e_{ij} = \sum_{k,l} \frac{1}{2} s_{kl}^{ij} (1 + \delta_{kl}) X_{kl}, \quad X_{ij} = \sum_{k,l} \frac{1}{2} c_{kl}^{ij} (1 + \delta_{kl}) e_{kl} \tag{11-11}$$

It is evident from Eq. (11-4) that under a coordinate transformation (R_{ij}) the combinations $1/2(1 + \delta_{ij}) e_{ij}$ transform like double products of coordinates

$$\frac{1}{2}(1 + \delta_{ij}) e_{ij}' = \sum_{k,l} R_{ik} R_{jl} \frac{1}{2}(1 + \delta_{kl}) e_{kl} \tag{11-12}$$

The stresses X_{ij}, on the other hand, transform strictly like second rank tensors

$$X_{ij}' = \sum_{k,l} R_{ik} R_{jl} X_{kl} \tag{11-13}$$

Substituting Eqs. (11-12) and (11-13) in Eq. (11-11) written for a primed coordinate system, and making use of the orthogonality properties of the transformation given in Eq. (3-6), we obtain the transformation rules

$$c_{mn}^{op'} = \sum_{i,j,k,l} R_{mi} R_{nj} R_{ok} R_{pl} c_{ij}^{kl} \tag{11-14}$$

and

$$\frac{1}{4}(1+\delta_{op})(1+\delta_{mn}) s_{mn}^{op'} = \sum_{\substack{i,j \\ k,l}} R_{mi} R_{nj} R_{ok} R_{pl} \frac{1}{4}(1+\delta_{ij})(1+\delta_{kl}) s_{ij}^{kl} \tag{11-15}$$

Hence the (c_{ij}^{kl}) transform like fourfold products of coordinates, while the (s_{ij}^{kl}) must in addition include a factor $\frac{1}{2}$ for each off-diagonal index pair. Neither the symmetries of the strains and stresses nor the pair symmetry of Eq. (11-9) are included in Eqs. (11-14) and (11-15). All such symmetries must be observed explicitly, whenever appropriate, to collect like terms in the transformed sums.

The same transformation rules apply to the two-index elastic parameters: (c_{ij}) transforms like a fourfold product of coordinates, but (s_{ij}) also includes a factor $\frac{1}{2}$ for every index greater than 3. If we employ the rules of Section 3-6 to simplify the formal manipulations in carrying out the tensor transformation, we can take into account the symmetry of the stresses and strains. According to the discussion of symmetric tensors in Section 4-1, the elastic tensors transform like fourfold products of the components of two distinct coordinate triples (x,y,z), each triple occurring twice. Using the same coordinate triple twice automatically reduces the number of independent quadratic combinations of its components from nine to six. As expected, there is then a one-to-one correspondence between the six products

$$x^2, y^2, z^2, yz, zx, xy$$

and each index running from 1 to 6.

As an example of these rules, let us find the transformed two-index elastic tensor components under the familiar coordinate transformation

$$(R_{ij}) = \begin{pmatrix} \cos\theta & \sin\theta & 0 \\ -\sin\theta & \cos\theta & 0 \\ 0 & 0 & 1 \end{pmatrix}$$

If the coordinate triples are (x_1, y_1, z_1) and (x_2, y_2, z_2) for the first and second indices, then, for example, the component c'_{14} is related to the unprimed components as follows.

$$c'_{14} \sim x'^2_1 y'_2 z'_2 = (x_1 \cos\theta + y_1 \sin\theta)^2(-x_2 \sin\theta + y_2 \cos\theta)z_2$$

$$= (x_1^2 \cos^2\theta + 2x_1 y_1 \cos\theta \sin\theta + y_1^2 \sin^2\theta)(-x_2 z_2 \sin\theta + y_2 z_2 \cos\theta)$$

By identifying the corresponding coordinate products on the right-hand side, we have

$$c'_{14} = -c_{15} \cos^2\theta \sin\theta + c_{14} \cos^3\theta - 2c_{65} \cos\theta \sin^2\theta + 2c_{64} \cos^2\theta \sin\theta$$
$$-c_{25} \sin^3\theta + c_{24} \cos\theta \sin^2\theta$$

The corresponding transformation for s'_{14} includes the factors $\frac{1}{2}$ according to the rule discussed in connection with Eq. (11-15)

$$\frac{1}{2}s'_{14} = -\frac{1}{2}s_{15} \cos^2\theta \sin\theta + \frac{1}{2}s_{14} \cos^3\theta - \frac{1}{2}s_{65} \cos\theta \sin^2\theta$$
$$+ \frac{1}{2}s_{64} \cos^2\theta \sin\theta - \frac{1}{2}s_{25} \sin^3\theta + \frac{1}{2}s_{24} \cos\theta \sin^2\theta$$

Other examples are given in Problem 11-4.

Once the transformation properties of the elastic parameters are established, the symmetry of the elastic tensors in various crystal systems can be determined by applying the rules of Section 3-7. Recapitulating briefly, this involves, first, using the generating elements of the group in question, as given in Appendix 2, to transform the parameters according to Eq. (11-14) or (11-15) to a new coordinate system. Second, since the new coordinate system is equivalent to the old one, the primed constants are equal to their unprimed values. Hence the transformation establishes relations between the various elements of the elastic tensor that lead to a form of this tensor compatible with the symmetry in question.

As an example, let us determine the scheme (c_{ij}) in cubic crystals. All cubic symmetries contain the threefold rotation B about the body diagonal, and at least a twofold rotation about one of the coordinate axes. The index symmetry imposed by these generating elements is

$$B: \quad 1 \to 2 \to 3 \to 1$$
$$\quad 4 \to 5 \to 6 \to 4$$

$$2: \quad 1 \to 1 \quad 4 \to -4$$
$$\quad 2 \to 2 \quad 5 \to -5$$
$$\quad 3 \to 3 \quad 6 \to 6$$

and the surviving scheme of constants is

$$(c_{ij}) = \begin{pmatrix} c_{11} & c_{12} & c_{12} & 0 & 0 & 0 \\ c_{12} & c_{11} & c_{12} & 0 & 0 & 0 \\ c_{12} & c_{12} & c_{11} & 0 & 0 & 0 \\ 0 & 0 & 0 & c_{44} & 0 & 0 \\ 0 & 0 & 0 & 0 & c_{44} & 0 \\ 0 & 0 & 0 & 0 & 0 & c_{44} \end{pmatrix} \qquad (11\text{-}16)$$

Hence a cubic crystal is characterized by three elastic parameters c_{11}, c_{12}, and c_{44} appearing in all entries equivalent under cubic symmetry.

The parameter schemes for all crystal symmetries are tabulated in the standard references. If necessary, they can easily be reconstructed by the methods just outlined. The resulting schemes of constants apply, of course, only in the system of axes for which the particular form of the generating elements holds. In all other coordinate systems the component entries are changed, and consist of linear combinations of the independent constants. In this connection it is sometimes of interest to know how many of these independent constants, and which ones, can be involved in some given component of the elastic tensor expressed with respect to an arbitrary coordinate system. Such questions can be answered by referring to the independent subspaces under which linear combinations of the elastic constants transform, as developed in Chapter 4. For example, Table A-3-3 gives an explicit list of such invariant subspaces for a fourth rank tensor of symmetry similar to that of (c_{ij}).

11.3. ISOTROPIC AND POLYCRYSTALLINE MEDIA

Tensor schemes in isotropic media have already been discussed in Chapter 4 from the point of view of tensor invariants. It follows from Eq. (4-4) that the elastic tensor in isotropic matter has only two independent constants. Hence one of the three constants for cubic symmetry appearing in Eq. (11-16) must be expressible in terms of the others. As shown in Problem 11-6 or 11-7, the required relation is

$$c_{44} = \frac{1}{2}(c_{11} - c_{12}) \tag{11-17}$$

To obtain the elastic constants of polycrystalline material in terms of the single crystal constants we have to make assumptions about the internal state of the material. If the crystal grains have random orientations, there are two extreme possibilities:

1. The local strain e_i is constant. This requires nonuniform stresses in order to compensate for the orientation-dependent anisotropy of the different grains. We then define an average stress and relate it to the uniform strain through Hooke's law

$$\langle X_i \rangle = \sum_j \langle c_{ij} \rangle e_j \tag{11-18}$$

2. The local stress X_i is constant. This results in nonuniform strain, but we can define an average strain by

$$\langle e_i \rangle = \sum_j \langle s_{ij} \rangle X_j \tag{11-19}$$

The averages indicated in Eqs. (11-18) and (11-19) are over the different grains having their crystal axes pointing in all directions. In practice, neither of the two formulations above can be strictly correct, and the average should really cover both terms in the products on the right-hand side. It can be shown, however, that Eqs. (11-18) and (11-19) set upper and lower bounds to the elastic parameters of polycrystalline material, and in that sense they are useful.

The averages indicated in Eqs. (11-18) and (11-19) are over differently oriented grains. Equivalently, they are averages of the elastic parameters of a single crystal defined with respect to an arbitrary coordinate system over all orientations of this coordinate system. In such a rotational average only those constants are nonvanishing that correspond to the linear rotational invariants of the tensor (c_{ij}). As discussed in Section 4-2, and in Problem 4-8, the elastic tensors have two invariants, which can be chosen to be of the form listed in Eq. (4-4). In terms of coordinate triples, the invariants are

$$(x_1^2 + y_1^2 + z_1^2) \cdot (x_2^2 + y_2^2 + z_2^2), \quad (x_1 x_2 + y_1 y_2 + z_1 z_2)^2$$

In terms of the respective elastic parameters, the invariants are

$$(c_{11} + c_{22} + c_{33}) + 2(c_{12} + c_{13} + c_{23})$$
$$(c_{11} + c_{22} + c_{33}) + 2(c_{44} + c_{55} + c_{66}) \tag{11-20}$$

or

$$(s_{11} + s_{22} + s_{33}) + 2(s_{12} + s_{13} + s_{23})$$
$$(s_{11} + s_{22} + s_{33}) + \frac{1}{2}(s_{44} + s_{55} + s_{66}) \tag{11-21}$$

The polycrystalline material is described by the isotropic tensor defined by Eqs. (11-16) and (11-17). Labeling the components of this tensor by C_{ij}, we can form its two invariants and equate them to Eq. (11-20). This yields the results

$$C_{11} = 1/15[3(c_{11} + c_{22} + c_{33}) + 2(c_{12} + c_{13} + c_{23}) + 4(c_{44} + c_{55} + c_{66})]$$
$$C_{12} = 1/15[(c_{11} + c_{22} + c_{33}) + 4(c_{12} + c_{13} + c_{23}) - 2(c_{44} + c_{55} + c_{66})] \tag{11-22}$$

Equation (11-2), valid for all symmetries, defines the elastic parameters of polycrystalline aggregates according to the average indicated in Eq. (11-18). In this approximation only 9 of the 21 elastic parameters enter into the behavior of the quasi-isotropic polycrystalline material.

The corresponding average in Eq. (11-19) yields the set of elastic constants

$$S'_{11} = 1/15[3(s_{11} + s_{22} + s_{33}) + 2(s_{12} + s_{13} + s_{23}) + (s_{44} + s_{55} + s_{66})]$$
$$S'_{12} = 1/15[(s_{11} + s_{22} + s_{33}) + 4(s_{12} + s_{13} + s_{23}) - \frac{1}{2}(s_{44} + s_{55} + s_{66})] \tag{11-23}$$

They are indicated by *primes* since they refer to a different averaging process and are not the inverses of Eq. (11-22).

11.4. DETERMINATION OF ELASTIC CONSTANTS BY STATIC METHODS

In a static experiment, prescribed stresses are applied to a sample subject to given boundary conditions, and we observe the resulting deformations. The limitations of the method arise from the requirement that observable deformations remain within the approximations of the infinitesimal theory discussed in this chapter.

11.4 Elasticity

Our principal aim in discussing such static experiments is the prediction of the maximum information obtainable in a particular experimental arrangement. We exemplify the approach by analyzing two of the simplest configurations in which the applied stresses are specified. If these stresses are uniform, the resulting deformation in terms of the stresses is obtained by combining Eqs. (11-4) and (11-11).

$$u_i = \sum_{j,k,l} \frac{1}{4}(1 + \delta_{ij})(1 + \delta_{kl})x_j s_{kl}^{ij} X_{kl} \qquad (11\text{-}24)$$

In the first example the same is a rectangular bar and is subjected to uniform pressure, as by immersion in a fluid. The applied stresses are

$$X_{11} = X_{22} = X_{33} = -P$$

and Eq. (11-24) reduces to

$$u_i = -P \sum_j \frac{1}{2}(1 + \delta_{ij})x_j(s_{11}^{ij} + s_{22}^{ij} + s_{33}^{ij}) \qquad (11\text{-}25)$$

Hence the distortion of the sample is governed by the six combinations of elastic constants (now written in two-index notion)

$$\Sigma_i = (s_{i1} + s_{i2} + s_{i3}), \qquad i = 1,\ldots,6 \qquad (11\text{-}26)$$

and measurement of the distortions yields the six Σ_i.

Since the applied stresses are isotropic, and since the shape of the sample is immaterial in obtaining Eq. (11-25), it is obvious that no new information can be obtained beyond the six constants Σ_i either by choice of coordinate axes or by choice of new sample shape. This is confirmed by inspection of Table A-3-3 where the six linear combinations of tensor components corresponding to Σ_i occur in only two subspaces, one of dimension five, the other of dimension one (see Problem 11-12).

In the second example, we consider a long thin cylindrical rod of arbitrary cross section under uniform tension T along its axis, as shown in Fig. 11-3. The usual experimental boundary conditions require that points originally on the axis of the rod remain on the x axis, Fig. 11-3b, so that the observed deformation includes a rotation.

The deformation satisfying these conditions is

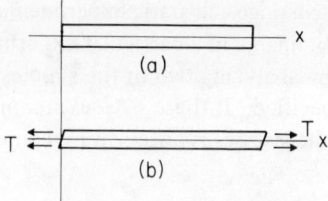

Fig. 11-3 Deformation of a long thin rod under uniform longitudinal tension T. (a) Undeformed sample. (b) Deformation leaving all points on the rod axis ($y = z = 0$) that were there initially.

$$\begin{pmatrix} u_1 \\ u_2 \\ u_3 \end{pmatrix} = T \begin{pmatrix} s_{11} & s_{16} & s_{15} \\ 0 & s_{12} & \frac{1}{2}s_{14} \\ 0 & \frac{1}{2}s_{14} & s_{13} \end{pmatrix} \begin{pmatrix} x \\ y \\ z \end{pmatrix} \qquad (11\text{-}27)$$

Hence the sample undergoes changes in both longitudinal and lateral dimensions. The cross section of the rod remains planar but inclines with respect to the x axis, and if $s_{12} \neq s_{13}$ it also changes shape.

A given orientation of the rod with respect to crystal axes yields the six constants s_{1i}, $i = 1,...,6$, expressed in a coordinate system whose x axis coincides with the axis of the rod. By choosing different directions of the rod axis relative to crystal symmetry axes, new sets of six constants are obtained. From inspection of Table A-3-3 it is evident that the six constants contained in Eq. (11-27) connect to all 21 elastic constants under suitable coordinate transformations. Hence in principle this experimental arrangement allows determining the full set of s_{ij} by using a sufficient number of independent orientations of the rod axis. The number needed is, of course, dependent on the number of independent constants as dictated by the symmetry group of the material.

11.5. ISOTHERMAL AND ADIABATIC CONSTANTS

The method of Section 11-4 is applicable when observable deformations still fall within the theory of infinitesimal elasticity leading to Eq. (11-27). An alternate method of determining elastic properties uses the ready detectability in most crystals of running or standing waves mechanically excited at the surfaces of the material. Here the principal measurement is concerned with a *velocity* rather than an amplitude, and difficulties arising from finite deformations can be minimized. On the other hand, the extraction of elastic constants from the experimental data presents its own difficulties, especially in highly anisotropic materials.

In comparing static elastic constants with those obtained ultrasonically, a correction enters because the latter are determined adiabatically, while a static experiment is isothermal. To establish the connection between isothermal and adiabatic properties, we assume that strains are caused by stresses X_i or by changes in temperature ΔT:

$$e_i = \alpha_i \Delta T + \sum_j s_{ij} X_j \tag{11-28}$$

where (α_i) is the tensor of *thermal expansion*, and (s_{ij}) is the *isothermal* elastic constants tensor. Since there exists a thermodynamic free energy density that is a function of T and X_i, the equation complementary to Eq. (11-28) is

$$\Delta S = \frac{1}{T_0} C_x \Delta T + \sum_i \alpha_i X_i \tag{11-29}$$

where ΔS is the change in entropy density, T_0 is the reference temperature, and C_x is the specific heat at *constant stress*.

By setting Eq. (11-29) equal to zero and eliminating ΔT in Eq. (11-28), the relation between adiabatic and isothermal elastic constants becomes

$$s_{ij}^{\text{ad}} - s_{ij}^{\text{iso}} = -T_0/C_x \, \alpha_i \alpha_j \tag{11-30}$$

Thus, the thermal expansion tensor as well as the specific heat at constant stress, usually taken at zero stress, enter into the connection. As shown later, measurements of the velocity of elastic waves give information about the inverse constants c_{ij}. The relation for these inverse constants corresponding to Eq. (11-30) is taken up in Problem 11-15.

11.6. ELASTIC WAVES

If the stresses X_{ij} acting on opposite sides of a volume element are not uniform, the mass contained in this volume will accelerate according to Newton's law

$$\rho \frac{\partial^2 u_i}{\partial t^2} = \sum_j \frac{\partial X_{ij}}{\partial x_j} \tag{11-31}$$

where ρ is the mass density of the crystal.

This equation can be transformed into a wave equation by rewriting the right-hand side in terms of the deformation components u_i. Using Eq. (11-11) and Eq. (11-4), we obtain

$$X_{ij} = \sum_{k,l} c_{kl}^{ij} \frac{\partial u_k}{\partial x_l} \tag{11-32}$$

where for the general time-dependent problems of Eq. (11-31) the elastic parameters must be taken as adiabatic. Hence the wave equation for elastic waves is

$$\rho \frac{\partial^2 u_i}{\partial t^2} = \sum_{j,k,l} c_{kl}^{ij} \frac{\partial^2 u_k}{\partial x_j \partial x_l} \tag{11-33}$$

We expect that in an unbounded medium this equation has basic solutions given by plane waves

$$u_i = U_i \exp[i(\mathbf{k} \cdot \mathbf{r} - \omega t)] \tag{11-34}$$

Inserting this solution in Eq. (11-33), we obtain, for the conditions on the wave amplitudes U_i

$$\sum_k \left(\sum_{j,l} c_{kl}^{ij} k_j k_l - \rho \omega^2 \delta_{ik} \right) U_k = 0 \tag{11-35}$$

The three homogeneous equations for U_i of Eq. (11-35) have a solution only if the secular equation of their coefficients is satisfied. This requirement leads to the familiar form of a determinantal equation for the propagation velocity $v = \omega/k$. If we write the propagation vector in terms of direction cosines as $\mathbf{k} = k(K_1, K_2, K_3)$, the secular determinant becomes

$$|\Gamma_{ik} - \rho v^2 \delta_{ik}| = 0 \tag{11-36}$$

where the coefficients Γ_{ik} are defined by

$$\Gamma_{ik} = \sum_{j,l} c_{kl}^{ij} K_j K_l \tag{11-37}$$

Equation (11-36) has three solutions for v^2; thus there are three phase velocities of elastic waves propagating in the direction of $\mathbf{K}$. The surface of phase velocities as a function of $\mathbf{K}$ has three sheets, one more than the corresponding surface for electromagnetic waves discussed in Chapters 9 and 10, because here there is no requirement that some field amplitudes be strictly transverse to $\mathbf{K}$.

Elasticity

The problem of reconstructing the scheme of elastic constants from measured phase velocities in various directions is in general formidable. Each direction **K** yields three velocities and hence, from Eq. (11-36), three combinations of the Γ_{ik}: linear, quadratic, and cubic in the Γ_{ik}, respectively. Using all possible directions of propagation, we can determine from the linear term at most six combinations of c_{ij}. Hence, use of the higher-order combinations is essential, and iteration procedures are required to find a self-consistent set of 21 elastic parameters based on measured velocities in at least seven independent directions. In practice, the problem is simplified in symmetric systems as much as possible by selecting propagation directions of high symmetry where Eq. (11-36) factors.

The mode amplitudes are the solutions of Eq. (11-35) once Eq. (11-36) is satisfied. If we define three factors a_i such that

$$\Gamma_{ij} = a_i a_j, \quad i \neq j \tag{11-38}$$

Eq. (11-35) can be rewritten

$$\sum_k (a_i a_k (1 - \delta_{ik}) - (\rho v^2 - \Gamma_{ii}) \delta_{ik}) U_k = 0 \tag{11-39}$$

or

$$U_i = \frac{a_i \sum_k a_k U_k}{\rho v^2 - \Gamma_{ii} + a_i^2} \tag{11-40}$$

The constant factor $\sum_k a_k U_k$ common to all three components of the mode amplitude can be absorbed in the arbitrary scale factor inherent in all solutions of this amplitude, so that the three components are in the ratio

$$U_1 : U_2 : U_3 = \frac{a_1}{\rho v^2 - \Gamma_{11} + a_1^2} : \frac{a_2}{\rho v^2 - \Gamma_{22} + a_2^2} : \frac{a_3}{\rho v^2 - \Gamma_{33} + a_3^2} \tag{11-41}$$

For a given direction **K**, the Γ_{ij} are common to all three velocities. The mode amplitudes of the three waves are orthogonal to each other, but generally do not correspond to either longitudinal or transverse waves.

The foregoing introduction to elastic waves leaves out the various other aspects of waves in anisotropic media that were touched on in Chapters 9 and 10 for electromagnetic waves. These include the distinction between energy flow and wave normal directions, the existence of singular directions where the velocity sheets intersect, triple refraction at crystal boundaries, mode mixing at interfaces, and so on. Analytic treatment of these problems is possible, though cumbersome.

Problems

11-1. (a) Verify that for a rigid rotation $\partial u_i/\partial x_j = -\partial u_j/\partial x_i$ by applying the transformation Eq. (3-20) for very small angles.
(b) Decompose the deformation in the xy plane

$$\begin{pmatrix} u_1 \\ u_2 \end{pmatrix} = \begin{pmatrix} 0.03 & 0.02 \\ 0.01 & -0.05 \end{pmatrix} \begin{pmatrix} x_1 \\ x_2 \end{pmatrix}$$

into a rotation and a pure deformation.

11-2. Determine the strains responsible for the deformation taking the following three coordinate points into new locations.

$(-2,0,3) \to (-2.032, -0.006, 2.948)$
$(6,5,-1) \to (6.079, 5.088, -1.004)$
$(3,-7,5) \to (2.989, -7.089, 4.888)$

How much of this information is redundant, and what is the minimum information needed to determine a complete set of strains?

11-3. Given a system of two nonvanishing strains e_{22}, e_{23}. Show, using Eq. (11-10), that the necessary conditions that the strain-free equilibrium be stable with respect to this deformation are given by

$$c_{22} > 0, \quad c_{44} > 0, \quad c_{22}c_{44} > c_{24}^2$$

11-4. Show that under the coordinate transformation in the example of Section 11-2 these results hold:
(a) $c'_{11} = \cos^4\theta\, c_{11} + 4\cos^3\theta \sin\theta\, c_{16} + 2\cos^2\theta \sin^2\theta\, (c_{12} + 2c_{66})$
$\quad + 4\cos\theta \sin^3\theta\, c_{26} + \sin^4\theta\, c_{22}$
(b) $s'_{23} = \sin^2\theta\, s_{13} - \sin\theta \cos\theta\, s_{63} + \cos^2\theta\, s_{23}$
(c) $s'_{44} = \sin^2\theta\, s_{55} - 2\sin\theta \cos\theta\, s_{54} + \cos^2\theta\, s_{44}$
(d) $c'_{66} = \cos^2\theta \sin^2\theta\, (c_{11} - 2c_{12} + c_{22}) - 2\cos\theta \sin\theta\, (\cos^2\theta - \sin^2\theta) \times (c_{16} - c_{26}) + (\cos^2\theta - \sin^2\theta)^2 c_{66}$

11-5. Use the direct inspection method to show that the elastic tensor in the tetragonal symmetry $(4/m)$ has the form

$$\begin{pmatrix} c_{11} & c_{12} & c_{13} & 0 & 0 & c_{16} \\ c_{12} & c_{11} & c_{13} & 0 & 0 & -c_{16} \\ c_{13} & c_{13} & c_{33} & 0 & 0 & 0 \\ 0 & 0 & 0 & c_{44} & 0 & 0 \\ 0 & 0 & 0 & 0 & c_{44} & 0 \\ c_{16} & -c_{16} & 0 & 0 & 0 & c_{66} \end{pmatrix}$$

11-6. (a) Show that under the generating element (6) the transformation of c_{66} is

$$c'_{66} = \frac{3}{16}(c_{11} + c_{22}) - \frac{3}{8}c_{12} - \frac{\sqrt{3}}{4}(c_{16} - c_{26}) + \frac{1}{4}c_{66}$$

(b) Argue that under sixfold symmetry around the z axis we must have the index symmetry $1 = 2$ in the base plane.

(c) Establish the relation

$$c_{66} = (c_{11} - c_{12})/2$$

for all hexagonal crystals.

(d) Show that the corresponding relation for s_{66} is

$$s_{66} = 2(s_{11} - s_{12})$$

(e) Show that these results are in agreement with the identity following Eq. (4-9) for isotropic matter.

11-7. (a) Show that a shear strain $e_6 = \eta$ transforms into the normal strains $e_1' = \frac{1}{2}\eta, e_2' = -\frac{1}{2}\eta$ in a coordinate system rotated $45°$ around the z axis with respect to the original frame of reference.

(b) Express the elastic energy density of a cubic crystal in both systems of coordinates, and show that if both frames are equivalent, then we must have

$$c_{44} = (c_{11} - c_{12})/2$$

11-8. Show that for the strain-free state of a cubic crystal to be stable with respect to all deformations, the elastic parameters must obey the inequalities

$$c_{11} > 0, \quad c_{11} > c_{12} > -\frac{1}{2}c_{11}, \quad c_{44} > 0$$

and similarly for the corresponding s_{ij}.

11-9. Show that the elastic constants and their inverses in a cubic crystal are related by

$$c_{11} = \frac{(s_{11} + s_{12})}{(s_{11} - s_{12})(s_{11} + 2s_{12})}, \quad c_{12} = \frac{-s_{12}}{(s_{11} - s_{12})(s_{11} + 2s_{12})}, \quad c_{44} = \frac{1}{s_{44}}$$

and conversely.

11-10. Calculate the elastic properties of a polycrystalline aggregate of a cubic crystal according to Eqs. (11-22) and (11-23), and show that

(a) $C_{11} = \dfrac{s_{11} + s_{12}}{(s_{11} - s_{12})(s_{11} + 2s_{12})} - \dfrac{2}{5}\dfrac{1}{(s_{11} - s_{12})}\dfrac{\Delta}{1 + \Delta}$

$$C_{12} = \frac{-s_{12}}{(s_{11} - s_{12})(s_{11} + 2s_{12})} + \frac{1}{5} \frac{1}{(s_{11} - s_{12})} \frac{\Delta}{1 + \Delta}$$

(b) $\quad C'_{11} = C_{11} - \frac{4}{25} \frac{1}{(s_{11} - s_{12})} \frac{\Delta^2}{(1 + \Delta)(1 + \frac{3}{5}\Delta)}$

$\quad C'_{12} = C_{12} + \frac{2}{25} \frac{1}{(s_{11} - s_{12})} \frac{\Delta^2}{(1 + \Delta)(1 + \frac{3}{5}\Delta)}$

where

$$\Delta = \frac{s_{44}}{2(s_{11} - s_{12})} - 1 = \frac{c_{11} - c_{12}}{2c_{44}} - 1$$

11-11. The traditional elastic parameters are defined in terms of specific applied stresses

Bulk modulus: $\quad B = \dfrac{\frac{1}{3}(X_1 + X_2 + X_3)}{e_1 + e_2 + e_3}, \quad X_1 = X_2 = X_3$ applied

Young's modulus: $\quad Y_i = X_i/e_i, \quad i = 1,2,3, \quad X_i$ applied

Shear modulus: $\quad S_i = X_i/e_i, \quad i = 4,5,6, \quad X_i$ applied

Poisson's ratio: $\quad \sigma_{ij} = -e_i/e_j, \quad i,j = 1,2,3, \quad X_j$ applied

(a) Express these quantities in terms of the elastic constants s_{ij}.
(b) Determine the elastic invariants in terms of B, Y_i, S_i, σ_{ij}.
(c) Show that in isotropic material there exist the relations

$$2\sigma = 1 - \frac{1}{3}\frac{Y}{B}, \quad S = \frac{3YB}{9B - Y}$$

(d) Show that in cubic crystals we have the restrictions

$$9 > \frac{Y}{B} > 0 \quad \text{and} \quad \frac{1}{2} > \sigma > -1$$

11-12. (a) Show that the first invariant given by Eq. (11-21) is equal to the inverse of the bulk modulus B.
(b) Show that using Eq. (11-26) this invariant can also be written

$$(\Sigma_1 + \Sigma_2 + \Sigma_3)$$

(c) Show that this invariant corresponds to the sum of the two one-dimensional invariant subspaces of Table A-3-3.
(d) Show that the components of the sum of the two five-dimensional subspaces can be expressed entirely in terms of the Σ_i of Eq.

(11-26). Argue that the six combinations constructed under (c) and (d) are a complete set of components only involving the Σ_i, and that therefore the Σ_i transform among themselves in an invariant six-dimensional space.

(Note: Observe that the entries in Table A-3-3 transform strictly like tensor components, while the s_{ij} do not.)

11-13. (a) Show that if three cylindrical rod samples in the example of Section 11-4 are chosen such that their axes coincide with each of the three crystal axes, the resulting deformations yield 15 independent elastic constants.

(b) Show that the remaining six elastic constants occur in six independent combinations in the deformation of a rod cut with its axis along the direction given by the transformation of Problem 3-12.

11-14. Specify the minimum number of cylindrical rod samples and their orientations in order to obtain a full set of elastic constants in the symmetry (4).

11-15. (a) Determine the free energy density giving rise to Eqs. (11-28) and (11-29) by starting from the expression for the first law of thermodynamics

$$dQ = T\,dS = dU - \sum_i X_i\,de_i$$

(b) Show that in terms of the independent variables $\Delta T, e_i$, the free energy density is

$$-\frac{1}{2}\frac{C_e}{T_0}\Delta T^2 + \sum_i \gamma_i e_i\,\Delta T + \frac{1}{2}\sum_{i,j} c_{ij}e_i e_j$$

where C_e is the specific heat at constant strain.

(c) Use (b) to show that

$$c_{ij}^{ad} - c_{ij}^{iso} = \frac{T_0}{C_e}\gamma_i \gamma_j$$

(d) Determine the relation between γ_i and α_i and show that

$$c_{ij}^{ad} = c_{ij}^{iso} + \frac{T_0}{C_e}\sum_{k,l} c_{ik}^{iso}\alpha_k c_{jl}^{iso}\alpha_l$$

11-16. (a) Show that for a cubic crystal, the first component of Eq. (11-33) can be written

$$\rho\frac{\partial^2 u_1}{\partial t^2} = c_{11}\frac{\partial^2 u_1}{\partial x^2} + c_{44}\left(\frac{\partial^2 u_1}{\partial y^2} + \frac{\partial^2 u_1}{\partial z^2}\right) + (c_{12} + c_{44})\left(\frac{\partial^2 u_2}{\partial x\,\partial y} - \frac{\partial^2 u_3}{\partial x\,\partial z}\right)$$

and that the other two components give the same equation under cyclic permutation of the coordinates.

(b) Show that a solution of the form $u_1 = u_1(x)$, $u_2 = u_3 = 0$ represents a longitudinal wave propagating along the x axis with a velocity $(c_{11}/\rho)^{1/2}$.

(c) Give the solution representing a transverse wave traveling along the y axis, and verify that its velocity is $(c_{44}/\rho)^{1/2}$.

(d) Show that a wave traveling in the (110) direction, with transverse amplitudes along ($-$110), obeys the same equation as that under (c) in a system of coordinates rotated $45°$ about the z axis, and use the result of Problem 11-7 to find the velocity of this wave

$$v = (\frac{1}{2}(c_{11} - c_{12})/\rho)^{1/2}.$$

11-17. (a) Show that in a coordinate system O' whose j axis is parallel to the direction of propagation of the elastic wave, Eq. (11-33) takes the simplified form

$$\frac{\partial^2 u'_i}{\partial t^2} = \sum_k c'^{ij}_{kj} \frac{\partial^2 u'_k}{\partial x'^2_j}$$

and leads to the secular equation

$$\left| c'^{ij}_{kj} - \rho v^2 \delta_{ik} \right| = 0$$

(b) Note that the primed elastic parameters are related to those in the crystal system of axes by a fourfold coordinate transformation, while the coefficients Γ_{ik} of Eq. (11-37) only involve a twofold transformation. Show that the secular determinant of (a) and that of Eq. (11-36) are related by a *similarity* transformation that leaves the determinant invariant.

11-18. (a) Derive the three phase velocities of elastic waves traveling in the (110) direction in a cubic crystal

$$\rho v^2 = c_{44}, \ c_{44} + \frac{c_{11} + c_{12}}{2}, \frac{c_{11} - c_{12}}{2}$$

(b) Determine the amplitudes of these three modes and show that one is longitudinal and two are transverse.

11-19. Show that the secular equation for elastic velocities in a hexagonal crystal can be factored into the form

$$[\frac{c_{11} - c_{12}}{2} + (c_{44} - \frac{c_{11} - c_{12}}{2}) K_3^2 - \rho v^2]$$

$$\times \left\{ [c_{11} + (c_{44} - c_{11}) K_3^2 - \rho v^2] [c_{44} + (c_{33} - c_{44}) K_3^2 - \rho v^2] \right.$$

$$\left. - K_3^2 (1 - K_3^2)(c_{13} + c_{44})^2 \right\} = 0$$

11-20. (a) Show that the mode of the velocity of the first factor of Problem 11-19 is a transverse wave with components in the xy plane.

(b) Determine the composition of the other two modes for $K_3 = 1$ and $K_3 = 0$.

(c) Show that for $0 < K_3 < 1$ the two modes of (b) are neither longitudinal nor transverse.

11-21. (a) Show that as a consequence of Eq. (11-37)

$$\Gamma_{ik} = \Gamma_{ki}$$

(b) Show that the linear combination of Γ_{ik} occurring in Eq. (11-36) is

$$\Gamma_{11} + \Gamma_{22} + \Gamma_{33}$$

and prove that by changing the direction of propagation **K**, at most six combinations of c_{ij} can be obtained from a knowledge of the linear combination of Γ_{ij}. Show that these six combinations involve 15 elastic constants.

11-22. (a) Show that compatibility of the left and right sides of Eq. (11-40) requires the identity

$$\sum_i \frac{a_i^2}{\rho v^2 - \Gamma_{ii} + a_i^2} = 1$$

(b) Show that any two modes with velocities v_1 and v_2 traveling in the same direction are orthogonal to each other, by using the result

$$[(\rho v_1^2 - \Gamma_{ii} + a_i^2)(\rho v_2^2 - \Gamma_{ii} + a_i^2)]^{-1} = [\rho(v_2^2 - v_1^2)]^{-1} \left\{ [\rho v_1^2 - \Gamma_{ii} + a_i^2]^{-1} - [\rho v_2^2 - \Gamma_{ii} + a_i^2]^{-1} \right\}$$

and the identity of (a).

(c) Check the results of Problem 11-18b by using Eq. (11-41).

Bibliography

J. F. Nye, *Physical Properties of Crystals*, Oxford Univ. Press, London and New York (1957), Chapters 5, 6.

S. Bhagavantam, *Crystal Symmetry and Physical Properties*, Academic Press, New York (1966), Chapter 11.

A. E. H. Love, *A Treatise on the Mathematical Theory of Elasticity*, Dover, New York (1944).

W. Voigt, *Lehrbuch der Kristallphysik*, B. Teubner, Berlin (1928).

F. I. Fedorov, *Theory of Elastic Waves in Crystals*, Plenum Press, New York (1968).

CHAPTER 12

Piezoelectricity

Some crystals subjected simultaneously to electric fields and to mechanical stresses show a response that differs from the sum of the reactions obtained when these forces are applied one at a time. Such *piezoelectric* materials couple electric and elastic behavior. This coupling gives rise to new effects. It also influences the electric and elastic properties studied separately in Chapters 5 and 11. In this chapter we take up the generalization of the earlier discussions that is required when piezoelectric coupling exists and we explore the new effects that it makes possible. More generally, the treatment of piezoelectricity also exemplifies the description of coupled phenomena that can be applied to any two or more of the properties analyzed separately in the preceding chapters.

12.1. THERMODYNAMICS OF PIEZOELECTRICITY

The piezoelectric coupling is a reversible phenomenon that is tied to the existence of a thermodynamic energy density. In a medium subject to changes in temperature ΔT, stresses X_i, and electric fields E_i, the first law of thermodynamics is

$$dU = T\,dS + \sum_{i=1}^{6} X_i\,de_i + \sum_{i=1}^{3} E_i\,dD_i \tag{12-1}$$

where U and S are the densities of internal energy and entropy. The free energy for the independent variables T, e_i, E_i follows from the Legendre transformation

$$F(T, e_i, E_i) = U - TS - \sum_i E_i D_i \tag{12-2}$$

12.1 Piezoelectricity

such that dF is a perfect differential

$$dF = -S\,dT + \sum_i X_i\,de_i - \sum_i D_i\,dE_i \tag{12-3}$$

To lowest order in changes away from the equilibrium configuration $T = T_0$, $e_i = 0$, $E_i = 0$, F must be a general quadratic form of the independent variables

$$F = -\frac{1}{2}\frac{C}{T_0}\Delta T^2 + \frac{1}{2}\sum_{i,j}c_{ij}e_ie_j - \frac{1}{2}\sum_{i,j}\epsilon_{ij}E_iE_j$$
$$+ \sum_i \gamma_i e_i \Delta T - \sum_i \delta_i E_i \Delta T - \sum_{i,j} f_{ij} E_i e_j \tag{12-4}$$

Equation (12-4) is a generalization of Eqs. (5-5) and (11-10) and of the free energy expression developed in Problem 11-15. Its new terms include the coupling coefficients δ_i and f_{ij} that connect the electric field to thermal and elastic effects. Since E_i and e_i have three and six components, respectively, the various sums in Eq. (12-4) involve different ranges of indices. In particular, the coefficient f_{ij} has its first index running over three and its second index running over six values. The dependent variables of Eq. (12-3) are given as first derivatives of F with respect to the independent variables. Hence Eq. (12-4) leads to the linear relations

$$\Delta S = \frac{C}{T_0}\Delta T - \sum_j \gamma_j e_j + \sum_j \delta_j E_j \tag{12-5}$$

$$D_i = \delta_i \Delta T + \sum_j f_{ij} e_j + \sum_j \epsilon_{ij} E_j \tag{12-6}$$

$$X_i = \gamma_i \Delta T + \sum_j c_{ij} e_j - \sum_j f_{ji} E_j \tag{12-7}$$

These equations identify the nature of the new coefficients. The first term in Eq. (12-6) predicts an electric polarization caused by changes in temperature. Hence (δ_i) describes a *pyroelectric* effect. Similarly, the second term in the same equation specifies a polarization caused by mechanical strains, so that (f_{ij}) is a *piezoelectric* tensor. The most important consequence of the thermodynamic approach is that the same sets of new coefficients are involved in the cross effects of Eqs. (12-5) and (12-7). Hence thermodynamics restricts the number of independent sets of coefficients required in the full phenomenological description.

The existence of the cross-terms in Eqs. (12-5), (12-6), and (12-7) also has an effect on the other coefficients that came from the purely thermal, electric, or elastic phenomena. All coefficients must now be defined carefully with respect to the other two variables. For example, the specific heat C of Eq. (12-5) is the *specific heat at zero strain and zero electric field*, and the elastic parameters c_{ij} of Eq. (12-7) refer to isothermal and field-free conditions. The ϵ_{ij} of Eq. (12-6) is the *isothermal and strain-free dielectric tensor*.

Coefficients valid under different constraints are obtainable from Eqs. (12-5), (12-6), (12-7) by inserting these constraints explicitly. For example, the adiabatic and strain-free dielectric tensor follows by setting $e_i = 0$ and $\Delta S = 0$ and eliminating ΔT

$$\epsilon_{ij}^{ad} = \epsilon_{ij}^{iso} - (T_0/C)\delta_i \delta_j \tag{12-8}$$

Similarly, the isothermal *stress-free* dielectric tensor is obtained by setting $\Delta T = 0$ and $X_i = 0$ and eliminating the strains e_i

$$D_i = \sum_j \left(\sum_{k,l} f_{ik} c_{kl}^{-1} f_{jl} + \epsilon_{ij} \right) E_j \tag{12-9}$$

Alternatively, we can define the new constants directly by starting from a free energy density of the proper independent variables. For example, if we take $\Delta T, X_i, E_i$ as the independent variables, the constitutive equations for D_i and e_i are

$$D_i = \delta_i' \Delta T + \sum_j h_{ij} X_j + \sum_j \epsilon_{ij}' E_j \tag{12-10}$$

$$e_i = \alpha_i \Delta T + \sum_j s_{ij} X_j + \sum_j h_{ji} E_j \tag{12-11}$$

The (ϵ_{ij}') is the dielectric constant tensor defined by Eq. (12-9), and the piezoelectric tensor (h_{ij}) is related to (f_{ij}) by

$$h_{ij} = \sum_k f_{ik} s_{kj} = \sum f_{ik} c_{kj}^{-1} \tag{12-12}$$

In constructing the coefficients valid under different constraint conditions it must be kept in mind that some coefficients may not always be realizable. For example, the elastically and electrically *free* state defined by $X_i = E_i = 0$ is realized by keeping the surface of a crystal stress free and at constant potential. On the other hand, while elastic clamping defined by $e_i = 0$ is easily achieved, electric clamping corresponding to $D_i = 0$ is difficult to arrange because the electric boundary conditions control only the normal component of **D**, although piezoelectric stresses induce both normal and tangential components.

12.2. TRANSFORMATION PROPERTIES OF PIEZOELECTRIC TENSORS

The basic piezoelectric relation contained in Eq. (12-6) is

$$D_i = \sum_j f_{ij} e_j \qquad (12\text{-}13)$$

and therefore the transformation properties of (f_{ij}) follow directly from those for D_i and e_j. For developing these rules it is convenient to rewrite f_{ij} in terms of the full three-index notation as f_i^{jk}, where the upper index pair refers to the strain. Because of the symmetry of the strains given by Eq. (11-3), f_i^{jk} is symmetric in the index pair

$$f_i^{jk} = f_i^{kj} \qquad (12\text{-}14)$$

In terms of this notation, Eq. (12-13) takes the form

$$D_i = \sum_{j,k} f_i^{jk} \frac{1}{2}(1 + \delta_{jk}) e_{jk} \qquad (12\text{-}15)$$

and direct substitution of the transformation law for vectors, Eq. (3-4), leads to the rule

$$f_i^{jk'} = \sum_{m,n,p} R_{ip} R_{jm} R_{kn} f_p^{mn} \qquad (12\text{-}16)$$

Hence the piezoelectric constants f_{ij} transform strictly like threefold coordinate products. On the other hand, the piezoelectric constants h_{ij} of Eq. (12-10) relating to the stresses follow the transformation rule

$$\frac{1}{2}(1 + \delta_{jk}) h_i^{jk'} = \sum_{m,n,p} R_{ip} R_{jm} R_{kn} \frac{1}{2}(1 + \delta_{mn}) h_p^{mn} \qquad (12\text{-}17)$$

which includes the factors $\frac{1}{2}$ whenever the stress index pair is off diagonal.

Equations (12-16) and (12-17) relate the piezoelectric coefficients of any two coordinate systems connected by the transformation (R_{ij}). In particular, they allow a complete determination of the (f_{ij}) and (h_{ij}) in all crystal systems. Thus, when the set of transformations (R_{ij}) refers to the symmetry operations of the crystal, Eq. (12-16) leads to a set of linear equations connecting various components of f_{ij} that establishes relations between these components. Using the approach of Chapter 3, we can readily determine the components' schemes in the different crystal symmetries. These schemes are well known and tabulated, so that the methods of Chapter 3 are primarily of interest to check the tabulated schemes or to transform the piezoelectric constants to other coordinate systems desirable because of the boundary conditions of a particular problem.

Since f_{ij} is a third rank polar tensor, it vanishes identically in all symmetry groups containing the element $(\overline{1})$. Similarly, crystal symmetries including rotation–inversions can show only limited piezoelectric interactions.

Piezomagnetism, which is the magnetic analogue of Eq. (12-13), is allowed in magnetic symmetry groups containing the elements (R_{ij}) discussed in Chapter 6 that combine spatial and time reversal symmetry. In addition, the replacement of **D** by **B** in Eq. (12-13) makes f_{ij}^* an axial tensor under inversion. With these restrictions, the formalism of piezomagnetic effects is very much the same as that developed here for piezoelectric interactions.

12.3. ELASTIC WAVES IN PIEZOELECTRIC CRYSTALS

A volume element in a piezoelectric crystal is subject to two kinds of forces. An electric field produces a polarization that distorts the content of a crystal cell. It acts on all the charges in the volume. An elastic stress acts on the surfaces enclosing the volume. But since the piezoelectric interaction given by Eq. (12-7) expresses the volume forces caused by an electric field in terms of stresses, all forces acting on a volume element are contained in the stresses X_{ij}. Consequently, the equations of motion of elastic media given by Eq. (11-31) remain valid without change in the presence of piezoelectric interactions. The effect of the electric fields is contained entirely in the modified form of Hooke's law given by Eq. (12-7), and in the accompanying expression for the electric displacement given by Eq. (12-6).

In order to apply these equations to wave propagation, which is adiabatic, we eliminate the terms proportional to ΔT through the condition $\Delta S = 0$ of Eq. (12-5). The adiabatic constants that result are those of Eq. (12-8) and of Problem 12-3. Just as in the discussion of waves in Chapter 11, their adiabatic character will be understood without introducing any explicit notation.

The modified form of Hooke's law deriving from Eq. (12-7) is, in three-index notation,

$$X_{ij} = \sum_{l,m} c_{lm}^{ij} \frac{\partial u_l}{\partial x_m} - \sum_n f_n^{ij} E_n \qquad (12\text{-}18)$$

and the accompanying equation for the electric displacement follows from Eq. (12-6)

$$D_p = \sum_{l,m} f_p^{lm} \frac{\partial u_l}{\partial x_m} + \sum_q \epsilon_{pq} E_q \qquad (12\text{-}19)$$

In addition to being subject to Eq. (11-31), piezoelectric waves also satisfy Maxwell's equations. For a wave of sufficiently low frequency the static approximation of these equations applies:

$$\nabla \times \mathbf{E} = 0, \qquad \nabla \cdot \mathbf{D} = 0 \qquad (12\text{-}20)$$

For a plane wave of phase form $\exp[i(\mathbf{k}\cdot\mathbf{r}-\omega t)]$, the first condition of Eq. (12-20) implies that the electric field accompanying the wave is *longitudinal*. Its vector amplitude is given by

$$\mathbf{E} = E^\circ \mathbf{k} \qquad (12\text{-}21)$$

The second condition of Eq. (12-20) establishes a relation between the electric field amplitude $\mathbf{E}$ and the deformation amplitude $\mathbf{U}$

$$0 = \sum_p k_p D_p = \sum_{l,m,p} (ik_m) k_p f_p^{lm} U_l + \sum_{p,q} k_p k_q \epsilon_{pq} E^\circ \qquad (12\text{-}22)$$

which determines the amplitude of the electric field

$$E^\circ = -\frac{\sum_{l,m,p}(ik_m)k_p f_p^{lm} U_l}{\sum_{p,q} k_p k_q \epsilon_{pq}} \qquad (12\text{-}23)$$

Using Eqs. (12-21) and (12-23) to eliminate the electric field in Eq. (12-18), we obtain the relation

$$X_{ij} = \sum_{l,m}(ik_m)\left[c_{lm}^{ij} + \frac{\sum_{n,p} k_n k_p f_n^{ij} f_p^{lm}}{\sum_{p,q} k_p k_q \epsilon_{pq}}\right] U_l \qquad (12\text{-}24)$$

which indicates that in wave propagation the net effect of the piezoelectric interaction is to modify the elastic parameters. As shown in the square bracket of Eq. (12-24), these modified parameters are dependent on the direction of propagation $\mathbf{k}$; hence, both the shape and the dimensions of the dispersion surfaces for elastic wave velocities are altered. The secular equation for these velocities is still given by Eq. (11-36), but the parameters Γ_{il} are modified to include the piezoelectric terms

$$\Gamma_{il} = \sum_{j,m} K_j K_m \left[c_{lm}^{ij} + \frac{\sum_{n,p} K_n K_p f_n^{ij} f_p^{lm}}{\sum_{p,q} K_p K_q \epsilon_{pq}} \right] \qquad (12\text{-}25)$$

where **K** designates the direction of **k**.

The amplitudes of the elastic deformation of a given wave are still defined by Eq. (11-41), and the specification of this wave is completed by giving the value of the accompanying longitudinal electric field according to Eqs. (12-21) and (12-23). The electric field is out of phase with the elastic deformation and in general it adds to the restoring force, so that the effective elastic parameters are increased as a result of piezoelectric coupling.

12.4. PIEZOELECTRIC EFFECTS IN ELECTROMAGNETIC WAVES

For the treatment of electromagnetic waves the static approximation of Maxwell's equations used in Eq. (12-20) must be replaced by the full electromagnetic equation of motion for **D**

$$\frac{\partial^2 D_p}{\partial t^2} = -\frac{1}{\mu_0} (\nabla \times (\nabla \times \mathbf{E}))_p \qquad (12\text{-}26)$$

Hence we now have two equations of motion. The first one, combining Eqs. (11-31) and (12-18), becomes an equation for the plane wave amplitudes

$$\rho v^2 U_i = \sum_{j,l,m} K_j K_m c_{lm}^{ij} U_l + i(v/\omega) \sum_{j,q} K_j f_q^{ij} E_q \qquad (12\text{-}27)$$

and the second, deriving from Eq. (12-26), is

$$v^2 D_p = \epsilon_0 c^2 \left(E_p - \sum_q K_p K_q E_q \right) \qquad (12\text{-}28)$$

These two equations, together with Eq. (12-19), determine a set of six linear homogeneous equations in the amplitudes U_i and E_i. Their combined solution yields five waves that correspond to three elastic and two electromagnetic modes, all modified by the piezoelectric coupling. Hence there are solutions with $v \sim v_{sound}$, and others with $v \sim v_{light}$. By assuming v of the proper magnitude, approximate solutions for both groups of waves can be found without going through a full and rigorous derivation.

Thus, if $v \sim v_{sound}$, the left-hand side of Eq. (12-28) is approximately equal to zero, and the vanishing of the right-hand side confirms that the electrical field accompanying elastic waves is longitudinal. On the other hand, if $v \sim v_{light}$, the first term on the right-hand side of Eq. (12-27) is negligible compared to the others. Hence there is a displacement accompanying the electric field given by

$$U_l = \frac{i}{\rho v \omega} \sum_{j,q} K_j f_q^{lj} E_q \qquad (12\text{-}29)$$

and this leads, by substitution in Eq. (12-19), to a modified dielectric constant

$$\epsilon'_{pq} = \epsilon_{pq} - (1\rho v^2) \sum_{j,l,m} K_j K_m f_p^{lm} f_q^{lj} \qquad (12\text{-}30)$$

This is the dielectric constant that enters into the formulas of crystal optics of Chapter 9 when they are applied to piezoelectric media. The correction term represents a direction-dependent softening of the dielectric constant. It is expected to be very small both because of the factor $1/v^2$ and because the piezoelectric coupling at very high frequencies is not strong.

12.5. OPTICAL PHONONS AND POLARITONS

The coupling between electromagnetic and elastic waves described in the preceding section leads only to a very small modification of both types of waves, essentially because their dispersion relations $\omega(\mathbf{k})$ are so very different. As shown schematically in Fig. 12-1, the only region of significant interaction

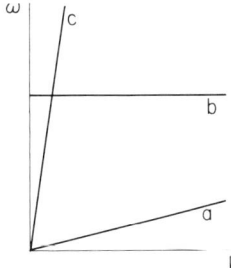

Fig. 12-1 Schematic dispersion relations for two branches of the elastic wave spectrum (curves a, b) and for electromagnetic waves (c). The ratio of the slopes for curves c and a is actually about 10^5.

between waves characterized by the dispersion curves of type a and c can be expected to occur at very low frequencies and long wavelengths. This is borne out by the results of Section 12-4. Figure 12-1 also includes an additional dispersion relation, curve b, which describes an entirely different branch of the spectrum of elastic waves in crystals. Such a branch of *optical phonons* can show a strong coupling to electromagnetic waves in the region where the dispersion curves b and c cross.

Optical phonon branches exist in all crystals that have more than one atom per cell. In such crystals the elastic distortions giving rise to wave propagation are of two basic types. In distortions of the first type, all the atoms of the crystal cell move in essentially the same phase. Hence the basic description of the distortion is the deformation of the unit cell, as given by Eq. (11-4), and the inertial response to applied forces involves the total mass of the unit cell. This center-of-mass motion leads to the wave equation given by Eq. (11-31), and to the *acoustic* branch of the elastic spectrum. The second type of distortion is concerned with displacement of the cell atoms relative to the center of mass. It involves a *reduced cell mass* and is described by an *internal displacement*. For n atoms per unit cell there are $(n-1)$ internal vector displacements, and each such displacement gives rise to three independent elastic waves. These waves retain a finite frequency as their wavelength becomes very large, and both their phase and group velocities are dependent on $\mathbf{k}$. They belong to the *optical* branches of the elastic spectrum.

If the internal displacement is accompanied by a redistribution of electrical charge in the unit cell, it can couple to an electric field. In the acoustic branch this coupling is given by the piezoelectric effect, and leads to the phenomena already described in this chapter. In the optical branches it provides the mechanism for strong interaction with electromagnetic waves.

The simplest macroscopic formulation of this interaction is in terms of internal displacement components w_i (normalized such as to absorb the reduced mass), linear restoring forces F_i, and electric fields E_i and D_i. The equations analogous to Eqs. (12-18) and (12-19) are

$$F_i = -\omega_0^2 w_i + fE_i \tag{12-31}$$

$$D_i = \quad fw_i + \epsilon E_i \tag{12-32}$$

where for simplicity we have assumed the solid to be *isotropic*.

These equations assume that the internal displacement at any one point in the solid is not coupled elastically to any other point. The only coupling between different parts of the solid is via the electric field. This is reflected in the fact that the equation of motion for w_i, given by

$$F_i = \partial^2 w_i / \partial t^2 \tag{12-33}$$

leads, in the absence of any coupling to electric fields, to a plane wave solution $\exp[i(\mathbf{k}\cdot\mathbf{r}-\omega t)]$ at a single frequency ω_0 regardless of $\mathbf{k}$.

$$f = 0, \quad w_i = W_i \exp[i(\mathbf{k}\cdot\mathbf{r}-\omega_0 t)] \tag{12-34}$$

The dispersion relation of this wave corresponds to curve b in Fig. 12-1.

When $f \neq 0$, all wave solutions must satisfy Eqs. (12-31), (12-32), (12-33), and (12-34). In terms of the wave amplitudes, this requires the conditions

$$W_i = \frac{f}{\omega_0^2 - \omega^2} E_i \tag{12-35}$$

$$D_i = \left(\epsilon + \frac{f^2}{\omega_0^2 - \omega^2}\right) E_i \tag{12-36}$$

and

$$\omega^2 D_i = (1/\mu_0)(k^2 E_i - k_i(\mathbf{E}\cdot\mathbf{k})) \tag{12-37}$$

Equation (12-35) specifies that $\mathbf{E}$ and $\mathbf{W}$ are always parallel. Equation (12-36) sets up a similar relation between $\mathbf{D}$ and $\mathbf{E}$, except at the frequency at which the frequency-dependent dielectric constant vanishes.

Hence, we expect two basic solutions. If $\mathbf{E}$ is parallel to $\mathbf{k}$, $\mathbf{D}$ must vanish, and thus we have a *longitudinal* wave given by

$$\omega^2 = \omega_0^2 + (f^2/\epsilon), \quad W_i^1 = (-\epsilon/f)E_i^1 \tag{12-38}$$

which represents a solution similar to that of Eq. (12-34) except that the frequency is pushed up because, just as in the piezoelectric case, the electric field stiffens the elastic constant.

If $\mathbf{E}$ is perpendicular to $\mathbf{k}$, so are $\mathbf{W}$ and $\mathbf{D}$, and there exist two degenerate *transverse* waves with the dispersion relation

$$\omega^2 \left(\epsilon + \frac{f^2}{\omega_0^2 - \omega^2}\right) = \frac{k^2}{\mu_0} \tag{12-39}$$

and with the various amplitudes at a given ω related according to Eqs. (12-35) and (12-36). At this point it is useful to introduce the frequency-dependent index of refraction $n^2(\omega) = \epsilon(\omega)/\epsilon_0$, where $\epsilon(\omega)$ is the dielectric constant of Eq. (12-36). By replacing the constants ϵ and f by the equivalent constants n_0 and n_∞ that give the limiting values of $n(\omega)$, we obtain, as the form for $n(\omega)$,

$$n^2(\omega) = n_\infty^2 \frac{(n_0/n_\infty)^2 \omega_0^2 - \omega^2}{\omega_0^2 - \omega^2} \tag{12-40}$$

In terms of these constants, the solution for ω^2 of Eq. (12-39) becomes

$$\omega^2 = (2n_\infty^2)^{-1} \left[n_0^2 \omega_0^2 + k^2 c^2 \pm \left[(n_0^2 \omega_0^2 + k^2 c^2)^2 - 4 n_\infty^2 \omega_0^2 k^2 c^2 \right]^{1/2} \right] \tag{12-41}$$

This dispersion relation is shown graphically in Fig. 12-2. As expected, it indicates strong coupling between the unperturbed dispersion curves of Fig. 12-1.

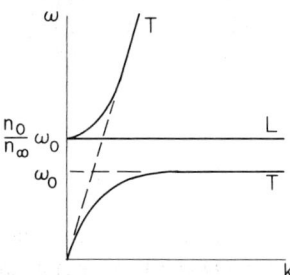

Fig. 12-2 Dispersion relation for longitudinal (L) and transverse (T) polariton waves, according to Eq. (12-41),

In particular, it exhibits a gap in the frequency spectrum such that no propagating modes exist in the range $1 < \omega/\omega_0 < n_0/n_\infty$. The range is identical with the splitting of the elastic modes in Eqs. (12-34) and (12-38) caused by their coupling to electric fields. Outside but near that gap, the electromagnetic waves that propagate contain a strong admixture of elastic excitation, as determined by Eq. (12-35). This coupling is particularly strong just below the gap where the group velocity of electromagnetic waves approaches the value characteristic of the purely elastic excitation. Some details of the division of electromagnetic and elastic energy transport in these modes are explored in Problem 12-15. Such modes are an example of *polariton waves*, waves that describe the normal modes of electromagnetic waves in interaction with various excitations of the crystal lattice due to the polar character of the crystal cell under deformation.

The gap predicted in our simple model is found in many crystals, generally in the infrared. Hence in such crystals electromagnetic waves of these and lower frequencies propagate primarily by means of the polarization accompanying the internal elastic displacements. Such waves can be expected to be lossy because the correct elastic energy density usually includes terms in the displacement higher than quadratic which lead to mode coupling and dissipation.

The discussion of polaritons presented here is intended primarily as a complement to the piezoelectric influences on elastic waves. It is obviously a major topic in its own right, and its detailed development must include such features as a proper formulation in anisotropic media, the inclusion of more realistic dispersion relations for optical phonons, and the coupling of polaritons to external waves through proper boundary conditions.

Problems

12-1. (a) Write down the free energy density leading to Eqs. (12-10) and (12-11).
 (b) Verify Eq. (12-12), and relate δ_i and δ_i'.

12-2. Show that the adiabatic stiffness constants at zero electric induction are given in terms of the coefficients of Eqs. (12-5) to (12-8) by

$$c_{ij}^{ad} = c_{ij} + (T_0/C)\gamma_i\gamma_j + \sum_{k,l}\left[(T_0/C)\gamma_i\delta_k + f_{ki}\right]\left(\epsilon_{kl}^{ad}\right)^{-1}\left[(T_0/C)\gamma_j\delta_l + f_{lj}\right]$$

12-3. Show that the adiabatic stiffness constants and piezoelectric constants belonging to the adiabatic dielectric constant of Eq. (12-8) are

$$c_{ij}^{ad} = c_{ij} + (T_0/C)\gamma_i\gamma_j, \qquad f_{ij}^{ad} = f_{ij} + (T_0/C)\delta_i\gamma_j$$

12-4. Verify Eq. (12-17).

12-5. Show that in the crystal symmetry $(\bar{4})$

$$(f_{ij}) = \begin{pmatrix} 0 & 0 & 0 & f_{14} & f_{15} & 0 \\ 0 & 0 & 0 & -f_{15} & f_{14} & 0 \\ f_{31} & -f_{31} & 0 & 0 & 0 & f_{36} \end{pmatrix}$$

12-6. Show that piezoelectricity is allowed in cubic crystals with symmetry (23) and $(\bar{4}3m)$, and specify the stresses required to produce polarization.

12-7. (a) Verify that the *piezomagnetic* tensor f_{ij}^* in the magnetic symmetry $(3\underline{2})$ has the form

$$(f_{ij}^*) = \begin{pmatrix} f_{11}^* & -f_{11}^* & 0 & 0 & f_{15}^* & 0 \\ 0 & 0 & 0 & f_{15}^* & 0 & -f_{11}^* \\ f_{31}^* & f_{31}^* & f_{33}^* & 0 & 0 & 0 \end{pmatrix}$$

(b) Construct the corresponding *piezoelectric* tensor in this system.

12-8. Show that although the piezomagnetic tensor is a third rank axial tensor, it has no linear invariant (Section 4-1), and that therefore piezomagnetism should vanish in polycrystalline materials.

12-9. (a) Show that both the numerator and denominator of Eq. (12-23) are invariant under a coordinate transformation (R_{ij}).
(b) Show that the piezoelectric correction term in Eq. (12-25) transforms under coordinate transformation like c^{ij}_{lm}.

12-10. Consider a piezoelectric crystal in the class $(6mm)$.
(a) Show that the three tensors entering into the propagation of elastic waves in this crystal are

$$(\epsilon_{ij}) = \begin{pmatrix} \epsilon_{11} & 0 & 0 \\ 0 & \epsilon_{11} & 0 \\ 0 & 0 & \epsilon_{33} \end{pmatrix}$$

$$(f_{ij}) = \begin{pmatrix} 0 & 0 & 0 & 0 & f_{24} & 0 \\ 0 & 0 & 0 & f_{24} & 0 & 0 \\ f_{31} & f_{31} & f_{33} & 0 & 0 & 0 \end{pmatrix}$$

$$(c_{ij}) = \begin{pmatrix} c_{11} & c_{12} & c_{13} & 0 & 0 & 0 \\ c_{12} & c_{11} & c_{13} & 0 & 0 & 0 \\ c_{13} & c_{12} & c_{33} & 0 & 0 & 0 \\ 0 & 0 & 0 & c_{44} & 0 & 0 \\ 0 & 0 & 0 & 0 & c_{44} & 0 \\ 0 & 0 & 0 & 0 & 0 & \frac{1}{2}(c_{11} - c_{12}) \end{pmatrix}$$

(b) Determine the Γ_{il} for $\mathbf{K} = (1,0,0)$ and $\mathbf{K} = (0,0,1)$ and show that in the first direction one of the transverse wave velocities is increased and in the second the longitudinal wave velocity is increased by piezoelectric coupling.
(c) Show that the modified waves are accompanied by an electric field.
(d) Specify a direction of $\mathbf{K}$ and a wave polarization that yields the third modified velocity needed to determine all three piezoelectric coefficients (see Problem 11-20).

12-11. (a) Use Eq. (12-30) to calculate the corrections to the dielectric tensor of Problem 12-10 if $\mathbf{K} = (K_1, 0, K_3)$.

(b) Derive the expression for the index of refraction for the two optical modes using the method of Problem 9-8.

12-12. Find the exact velocities and propagation modes of elastic and electromagnetic waves in a cubic piezoelectric crystal for $\mathbf{K} = (1,0,0)$, and show that only the transverse waves are coupled.

12-13. Determine the phase and group velocities of the optical phonons whose dispersion relation is shown in Fig. 12-1.

12-14. (a) Show that starting with the expression for the first law,

$$dU = \sum_i E_i dD_i - \sum_i F_i dw_i,$$

we can define a free energy density

$$F(E_i, w_i) = U - \Sigma E_i D_i$$

such that

$$F_i = -\partial F/\partial w_i, \qquad D_i = -\partial F/\partial E_i$$

and that a form

$$F = \frac{1}{2}\omega_0^2 w^2 - f\mathbf{w} \cdot \mathbf{E} - \frac{1}{2}\epsilon E^2$$

yields Eqs. (12-31) and (12-32).

(b) Show that in the static approximation of Maxwell's equations ($\nabla \times \mathbf{E} = 0$, $\nabla \cdot \mathbf{D} = 0$) the allowed internal displacement waves have a longitudinal branch given by Eq. (12-38) and a degenerate transverse branch given by Eq. (12-34).

(c) Calculate the free energy densities associated with these two modes and show that in each case

$$\left\langle \frac{1}{2}\dot{w}^2 \right\rangle_{time} = \langle F \rangle_{time}$$

12-15. (a) Show that the total internal energy density (including "kinetic" contributions) can be written

$$U = \frac{1}{2}(\mu_0 H^2 + \epsilon_0 n_\infty^2 E^2) + \frac{1}{2}(\dot{w}^2 + \omega_0^2 w^2)$$

and that this expression leads to the proper rate of change of electromagnetic energy

$$\frac{\partial U}{\partial t} = \mathbf{H} \cdot \frac{\partial \mathbf{B}}{\partial t} + \mathbf{E} \cdot \frac{\partial \mathbf{D}}{\partial t}$$

(b) Show that the time-averaged fraction of the energy density that resides in the internal displacement is given by

$$\frac{<\frac{1}{2}\dot{w}^2 + \omega_0^2 w^2>}{<U>} = \frac{[1 + (\omega^2/\omega_0^2)](n_0^2 - n_\infty^2)}{2(n_0^2 - n_\infty^2) + n_\infty^2 [1 - (\omega^2/\omega_0^2)]^2}$$

and evaluate the range of frequencies for which this fraction is above $\frac{1}{2}$.

12-16. Extend the theory of polaritons of Section 15-5 to a crystal of tetragonal symmetry.

Bibliography

J. F. Nye, *Physical Properties of Crystals*, Oxford Univ. Press, London (1957), Chapter 7.

W. P. Mason, *Crystal Physics of Interaction Processes*, Academic Press, New York (1966), Chapter 7.

W. G. Cady, *Piezoelectricity*, McGraw-Hill, New York (1946).

M. Born and K. Huang, *Dynamical Theory of Crystal Lattices*, Oxford Univ. Press, London (1954), Section 7, 8.

S. Bhagavantam, *Crystal Symmetry and Physical Properties*, Academic Press, New York (1966), Chapter 14.

V. M. Agranovich and V. L. Ginzburg, *Spatial Dispersion in Crystal Optics and the Theory of Excitons*, Interscience, New York (1966).

C. Kittel, *Introduction to Solid State Physics*, 4th ed., Wiley, New York (1971), Chapter 5.

CHAPTER 13

Formulation of Higher-Order Interactions

In the preceding chapters we have developed the basic formulation of anisotropic effects in various branches of macroscopic physics. We have also included examples of the simultaneous presence of more than one anisotropic interaction. It is clear that any extension of this formulation to a larger number of interactions involving the same basic effects falls into the pattern we have been following. Once a new interaction is postulated and its intrinsic symmetry is known, the phenomenological tensor scheme describing the interaction in different crystal symmetries can be worked out. If the new interaction is small in the sense of the next term of a converging power series, it modifies the fields and currents or forces and displacements developed so far by straightforward perturbation treatment. In general, the symmetry of the response is reduced, and degeneracies are lifted, by such added higher-order interactions.

In this chapter we deal with miscellaneous comments and problems in constructing and analyzing typical tensors of higher rank. While well-established mathematical methods exist for handling all such problems, we want to stress some of the shortcuts available for minimizing the formal work. Examples are given of tensors up to sixth rank, and the methods and tables developed in Chapter 4 are offered to extend the treatment to tensors of still higher order, if needed. In addition, we treat in further detail isotropic and axially symmetric media and the properties of polycrystalline aggregates of anisotropic crystallites.

13.1. VARIOUS FOURTH RANK TENSORS

The only fourth rank tensor discussed so far describes the linear relation of Hooke's law between stresses and strains

$$X_{ij} = \sum_{k,l} c_{kl}^{ij} e_{kl} \qquad (13\text{-}1)$$

Because of the three symmetries

$$(ij) \longleftrightarrow (ji), \quad (kl) \longleftrightarrow (lk), \quad (ij) \longleftrightarrow (kl) \tag{13-2}$$

the elasticity tensor has at most 21 independent components. Since this tensor is known for all crystal symmetries, it forms a good starting point for constructing all fourth rank tensors of equal or higher symmetry.

For example, the fourth-order *magnetic anisotropy energy density* in ferromagnetic single crystals is given by

$$U = \sum_{i,j,k,l} K_{ijkl} \alpha_i \alpha_j \alpha_k \alpha_l \tag{13-3}$$

where the α_i are the direction cosines of the magnetization density with respect to crystal axes. Since the order of the α_i's in Eq. (13-3) is immaterial, (K_{ijkl}) is a totally symmetric fourth rank tensor. In addition to the symmetries of Eq. (13-2) it obeys the rule

$$(ij)(kl) \longleftrightarrow (il)(kj) \tag{13-4}$$

The number of independent components is equal to the number of independent products in the expansion $(x + y + z)^4$. There are 15 such products. The same number follows from the group character $\chi_{(3 \times 3 \times 3 \times 3)_s}(\phi)$ of Table A-4-2 for $\phi = 0$, as already discussed in Chapter 4. Hence, using the known scheme of coefficients of (c_{ij}) in any crystal symmetry, the tensor (K_{ijkl}) can be readily constructed merely by imposing on this scheme the additional requirement of Eq. (13-4). Since (K_{ijkl}) is taken to transform exactly like a fourfold product of coordinates, it is important to use the scheme of the (c_{ij}) rather than of the (s_{ij}) in carrying out the reduction.

Fourth rank tensors that obey only the first two of the three symmetry relations of Eq. (13-2) have up to 36 independent components. They represent a less symmetric scheme than the elastic parameters of Eq. (13-1), and hence must be constructed independently. This can be done, for example, by using the methods of Chapter 3, or those of Chapter 4 in conjunction with Table A-3-3. These schemes, however, are readily available in standard references. An example of such interaction is given by the *photoelastic tensor* (P_{kl}^{ij}) in the relation

$$\Delta \epsilon_{ij} = \sum_{k,l} P_{kl}^{ij} X_{kl} \tag{13-5}$$

which describes the change of the dielectric constant under stress. Since both (ϵ_{ij}) and (X_{kl}) are symmetric, but there is no index pair symmetry between them, Eq. (13-5) defines a tensor with 36 independent coefficients. If the pair symmetry $(ij) = (kl)$ is imposed because of crystal symmetry, then the scheme (P_{ij}) reverts back to that of (c_{ij}). For instance, the threefold axis in cubic crystals automatically connects entries on both sides of the diagonal: $P_{12} = P_{23} = P_{31}$, and if a further symmetry requirement equalizes the coefficients on one side of the diagonal, as in the systems $(\bar{4}3m)$, (43), $(m3m)$, then (P_{ij}) follows the same scheme as (c_{ij}). In general, though, (P_{ij}) has additional and independent entries.

The scheme of (P_{ij}) also applies directly to other important interactions. For example, the *quadratic magnetoconductivity* introduced in Eq. (7-39) describes a change in conductivity depending on B^2

$$\Delta\sigma_{ij} = \sum_{k,l} A_{ijkl} B_k B_l \qquad (13\text{-}6)$$

The Tensor (A_{ijkl}) is fully equivalent to (P^{ij}_{kl}) and the coefficient schemes for the latter apply immediately.

Other fourth rank tensor schemes can be similarly related. A more unusual example is the *third-order Hall effect* given by the next term in the expansion of Eq. (7-40)

$$\Delta a_{ij} = \sum_{k,l,m} T_{ijklm} B_k B_l B_m \qquad (13\text{-}7)$$

Since by definition $a_{ij} = -a_{ji}$, the first index pair represents a single (axial) vector component. With this identification of the pair (ij) with a single index, $(T_{(ij)klm})$ is a fourth rank tensor. The symmetry differs appreciably from those discussed previously. It is totally symmetric in the last three indices, but there is no symmetry between these and the first index (ij). Hence the total number of independent components is $3 \times$ [number of independent products in $(x+y+z)^3$] $= 30$. In addition, considered as a fourth rank tensor, it is polar. The individual coefficient schemes must be derived by direct construction or by reduction from the full fourth rank tensor scheme or other applicable schemes of related but lower symmetry. In this particular case it is easiest to proceed by direct construction, since there is an immediate reduction from 81 to 30 components. The direct inspection method of Chapter 3 works very rapidly for a threefold totally symmetry product.

13.2. FIFTH AND SIXTH RANK TENSORS

The general fifth and sixth rank tensors have 243 and 729 components, respectively. It is obviously of interest to introduce at once as much symmetry as possible when discussing such tensors in order to keep the bookkeeping manageable. As a matter of fact, a complete listing of all their components in all crystal systems has been given by Fieschi and Fumi (1953). In principle, any more symmetric tensors can be deduced from these tables by introducing the additional symmetry.

In practice, only relatively few of the more symmetric tensors of these ranks have found applications so far. A typical fifth rank tensor describes the *electric field dependence of the elastic constants*.

$$c_{ij}(\mathbf{E}) = c_{ij} + \sum_k d_{ijk} E_k \tag{13-8}$$

The tensor (d_{ijk}) is symmetric in the first two indices running from 1 to 6, so that its most general scheme contains 63 independent coefficients. It is a polar fifth rank tensor which vanishes identically in all crystal symmetry groups containing the inversion $\bar{1}$ as a symmetry element. When it exists, its structure is not just a simple product of the elastic constants and the components of a vector allowed in the given crystal symmetry, because while each of these two factors may go separately into its negative under some symmetry operation, their product will not, and can therefore exist. This is exemplified in Problem 13-8.

If we replace the electric field as an expansion parameter in Eq. (13-8) by the wave vector direction **K** of elastic waves, we obtain a tensor describing acoustical rotatory activity. The symmetry of this tensor is constrained by an equation similar to Eq. (9-7) for the corresponding optical case. Additional examples of fifth rank tensors are taken up in Problems 13-9 and 13-10.

A sixth rank tensor of common interest describes the *third-order elastic coefficients* defined by the next terms of Hooke's law after Eq. (13-1)

$$X_{ij} = \sum_{\substack{k,l \\ m,n}} c^{ij}_{klmn} e_{kl} e_{mn} \tag{13-9}$$

The intrinsic symmetry of these constants derives from the existence of an energy density

$$U = \frac{1}{6} \sum_{\substack{i,j,k \\ l,m,n}} c^{ij}_{klmn} e_{ij} e_{kl} e_{mn} \tag{13-10}$$

It requires that all three pairs of symmetric indices be interchangeable with each other. This leads to a total of $(8!/5!3!) = 56$ possible independent coefficients, which correspond to all the distinct products of the expression $(\Sigma_i e_i)^3$. Some properties of this tensor are discussed in Section 13-5.

13.2   Formulation of Higher-Order Interactions

Other sixth rank tensors can be related to the behavior of the elastic parameters under the influence of other fields. For example, the dependence of the elastic constants of Eq. (13-1) on the *direction of magnetization* in a magnetic crystal can be written

$$\Delta c_{kl}^{ij} = \sum_{m,n} R_{klmn}^{ij} \alpha_m \alpha_n \qquad (13\text{-}11)$$

Here the symmetry of the index pairs (ij) and (kl) is that of the elastic parameters given by Eq. (13-2), while the last two indices are also symmetric, $(mn) = (nm)$, but are otherwise unrelated to the other indices. The total number of independent coefficients is therefore $21 \times 6 = 126$. The tensor is more general than that of Eq. (13-9) and, in fact, the latter can be obtained from the former by introducing the additional pair symmetry.

Similarly, *second-order magnetostriction* describes the fourth-order dependence of stress on the direction of magnetization

$$X_{ij} = \sum_{k,l,n,m} N_{klmn}^{ij} \alpha_k \alpha_l \alpha_m \alpha_n \qquad (13\text{-}12)$$

This tensor is the combination of a second rank symmetric tensor and a totally symmetric fourth rank tensor as described by Eq. (13-3). Hence it has a maximum of $6 \times 15 = 90$ components. It is derivable from the tensor of Eq. (13-11) by introducing the additional symmetry of Eq. (13-4).

Finally, the next-order term in the magnetic anisotropy energy of Eq. (13-3) is given by

$$U = \sum_{\substack{i,j,k \\ l,m,n}} K_{ijklmn} \alpha_i \alpha_j \alpha_k \alpha_l \alpha_m \alpha_n \qquad (13\text{-}13)$$

It involves a totally symmetric sixth rank tensor, of 28 independent coefficients, which is related to the other sixth rank tensors by introducing the maximum symmetry among the indices.

In all of these examples it is usually preferable to construct the tensor scheme directly for the symmetry group of interest rather than to work out the most general case first. The methods of Chapter 3 and 4, together with the material contained in the Appendices, allow a rather rapid construction in most cases.

Tensors of yet higher rank can be constructed as needed, using the same procedures. An obvious example of an eighth rank tensor is given by the fourth-order elastic parameters representing the next term in the expansion following Eqs. (11-10) and (13-10).

$$U = (1/24) \sum_{\substack{i,j \\ k,l}} c_{ijkl} e_i e_j e_k e_l \qquad (13\text{-}14)$$

where we have used the one-index notation for the symmetric index pairs describing strains. This tensor has only 126 independent components, rather than the 3^8 of a general eighth rank tensor, because of the total symmetry of the four indices.

13.3. ISOTROPIC AND CYLINDRICAL SYMMETRY

The schemes of higher-order tensors in isotropic media are a special case of the tensors in single crystals. Their construction requires, first, determination of the number of independent coefficients we may expect, and second, a convenient procedure for an explicit identification of all nonvanishing coefficients.

The number of independent coefficients, as already described in Chapter 4, is equal to the number of independent linear invariants of the tensor in question. This number may be obtained by the decomposition of the tensor character into its irreducible representations (Section 4-4), or it can also be derived from the generalization of Eq. (4-31) which applies to the infinite group of all three-dimensional rotations ϕ

$$n_{\text{iso}} = (1/2\pi) \int_0^{2\pi} \chi(\phi)(1 - \cos\phi) \, d\phi \qquad (13\text{-}15)$$

Rotation inversions are handled separately, depending on the polar or axial character of the tensor, in accord with Eq. (4-29).

Explicit construction of the tensor scheme is usually done most efficiently by first developing the tensor in the most symmetric finite group, the full cubic symmetry, and then applying the direct inspection method to an additional general rotation, such as 45°, in order to obtain the further relations between the constants when the cubic crystal becomes isotropic. Simple examples of this procedure applied to the elastic constants are given in Problems 11-6 and 11-7. A more complex isotropic tensor is worked out in Section 13-5.

In a medium of cylindrical symmetry with respect to some axis, the number of independent coefficients is given by

$$n_{\text{cyl}} = (1/2\pi) \int_0^{2\pi} \chi(\phi) \, d\phi \qquad (13\text{-}16)$$

where $\chi(\phi)$ is the tensor character for arbitrary rotations. In contrast to the isotropic medium, an axially symmetric medium is not necessarily more symmetric than all of the finite groups of crystal symmetry, so that the number of constants predicted by Eq. (13-16) is often larger than that applying in the highly symmetric groups. For example, for a second rank symmetric tensor like the dielectric constant, which has a character $\chi(\phi) = 2\cos\phi + 4\cos^2\phi)$, Eq. (13-16) gives $n_{cyl} = 2$. This agrees with the results derived in Section 5-2 for rhombohedral, tetragonal, and hexagonal media. It is larger than the number of dielectric constants in cubic crystals, primarily because these crystals must be "isotropic" with respect to more than one axis of rotation.

For tensors up to and including fifth rank, the scheme of coefficients of an axially symmetric tensor is identical to that of the tensor in the group (6). The sixfold axis of this group requires isotropy in the plane normal to the axis for all invariants made up of coordinate products of power lower than six. Cylindrically symmetric tensors of higher rank are most directly constructed from those known for the group (6), or for the group (4) by applying the direct inspection method to an additional rotation, such as $45°$, in order to obtain the further relations between the coefficients of the finite symmetry group for assuring complete axial isotropy.

13.4. POLYCRYSTALLINE MEDIA

The common isotropic materials are usually polycrystalline aggregates composed of randomly oriented crystallites. The connection between the properties of such an isotropic solid and those of its single crystallites is normally complex. It depends on the boundary conditions of the forces and displacements or fields and currents at all interfaces between the crystallites, and on the spatial average over these locally nonuniform disturbances. If we can justify that this average is merely the mean of the single crystal interaction for all directions, the problem can be handled very simply. As already discussed in Sections 4-2 and 11-3, the only tensor quantities that survive in such a mean are the linear rotational invariants of the tensor. Once these invariants are known in terms of single crystal constants, they can be equated to the corresponding expressions in terms of the constants of the isotropic material. This suffices to determine fully the equivalent constants of the polycrystalline medium.

The most important step in this procedure is the identification of the linear invariants of a particular tensor. Following the approach of Chapter 4, we can construct the invariants most directly by forming scalars from the vectors $\mathbf{V}_i$, $\mathbf{V}_j$, ... associated with the tensor indices $i, j, \ldots$, where the vectors are distinct if the indices are independent but can be chosen to be the same if the indices are interchangeable. The number of independent scalars formed from these vectors is given by Eq. (13-15). Within this number, there is a certain arbitrariness of the way in which the invariants are chosen. The particular choice depends on the simplicity of formulation, the need for orthogonality, and so on.

As an example, let us develop the invariants for the elastic coefficient tensors of different orders given by Eqs. (13-1), (13-9), and (13-14). The tensor c_{kl}^{ij} is represented by two vectors $\mathbf{V}_1$ and $\mathbf{V}_2$, each appearing twice, and these have the two independent invariants

$$I_1(2) = V_1^2 V_2^2, \qquad I_2(2) = (\mathbf{V}_1 \cdot \mathbf{V}_2)^2 \qquad (13\text{-}17)$$

In this notation the number 2 identifies the invariants as belonging to the second-order elastic coefficients. Symmetry *between* index pairs has to be introduced explicitly in Eq. (13-17). This result has already been discussed in Chapter 4 and Section 11-3.

The three invariants of the third-order elastic constants are constructed from three vectors $\mathbf{V}_1, \mathbf{V}_2, \mathbf{V}_3$, each taken twice. They can be chosen to be

$$I_1(3) = V_1^2 V_2^2 V_3^2$$

$$I_2(3) = (\mathbf{V}_1 \cdot \mathbf{V}_2)(\mathbf{V}_2 \cdot \mathbf{V}_3)(\mathbf{V}_3 \cdot \mathbf{V}_1) \qquad (13\text{-}18)$$

$$I_3(3) = \frac{1}{3}\left\{V_1^2(\mathbf{V}_2 \cdot \mathbf{V}_3)^2 + V_2^2(\mathbf{V}_3 \cdot \mathbf{V}_1)^2 + V_3^2(\mathbf{V}_1 \cdot \mathbf{V}_2)^2\right\}$$

where again symmetry between index pairs must be included explicitly after these expressions are expanded.

Similarly, the four invariants of the fourth-order elastic parameters are given by the set

$$I_1(4) = V_1^2 V_2^2 V_3^2 V_4^2, \qquad I_2(4) = (\mathbf{V}_1 \cdot \mathbf{V}_2)^2 (\mathbf{V}_3 \cdot \mathbf{V}_4)^2$$

$$(13\text{-}19)$$

$$I_3(4) = V_1^2 V_2^2 (\mathbf{V}_3 \cdot \mathbf{V}_4)^2, \qquad I_4(4) = (\mathbf{V}_1 \cdot \mathbf{V}_2)(\mathbf{V}_2 \cdot \mathbf{V}_3)(\mathbf{V}_3 \cdot \mathbf{V}_4)(\mathbf{V}_4 \cdot \mathbf{V}_1)$$

In analogy with Eq. (13-18), the invariants $I_2(4)$ and $I_3(4)$ could have been written in a form fully symmetric in the four vectors. Because of the symmetry between pairs, however, the final form of the expanded invariant will contain the full symmetry in either formulation.

Once the invariants are known, the procedure for connecting the single crystal constants to those of the polycrystalline medium is straightforward. It has already been carried out for Eq. (13-1) in Section 11-3. The detailed construction applying to Eq. (13-9) is taken up in Section 13-5, and the corresponding procedure for the fourth-order elastic parameters is outlined in Problem 13-17.

13.5. THIRD-ORDER ELASTIC COEFFICIENTS IN CUBIC, ISOTROPIC, AND POLYCRYSTALLINE MEDIA

In this section we demonstrate the procedure discussed in Section 13-4 for the third-order elastic coefficients of Eq. (13-9). To determine the *number* of coefficients to be expected in various symmetries we construct the character of the transformation of the tensor c_{ijk} ($i, j, k = 1, ..., 6$). Since all three indices are interchangeable, the character is that of a triple totally symmetric product of a symmetric second rank tensor. According to the second line of Table A-4-4, this product is given by

$$\frac{1}{3}\chi_t(3\phi) + \frac{1}{2}\chi_t(2\phi)\chi_t(\phi) + \frac{1}{6}\chi_t^3(\phi) \tag{13-20}$$

and if $\chi_t(\phi)$ is that of a symmetric second rank tensor

$$\chi_t(\phi) = 2\cos\phi + 4\cos^2\phi \tag{13-21}$$

the substitution of Eq. (13-21) in Eq. (13-20) leads to the formula for the tensor character

$$\chi(\phi) = 16\cos^2\phi - 8\cos^3\phi - 48\cos^4\phi + 32\cos^5\phi + 64\cos^6\phi \tag{13-22}$$

Equation (13-22) agrees with the next to last entry of Table A-4-2, where the character of the triple totally symmetric product of a symmetric second rank tensor has already been tabulated. The maximum number of independent coefficients of this tensor is given by the character for $\phi = 0$. $\chi(0) = 56$, in agreement with the discussion following Eq. (13-10).

The number of invariants is obtained by using Eq. (13-22) in Eq. (13-15). Alternatively, it follows from the decomposition of the character of Eq. (13-32) into its irreducible components according to the procedure of Section 4-4. The result is

$$\chi(\phi) = 3\chi_1 + 3\chi_5 + \chi_7 + 2\chi_9 + \chi_{13} \tag{13-23}$$

indicating that the totally symmetric one-dimensional representation χ_1 occurs three times to give three invariants. The sum of the products of the dimensions of the irreducible representations and the number of times they occur, of course, adds up to 56. Equation (13-23) verifies the next to last entry of Table A-4-3, where such irreducible decompositions are listed for a variety of tensor characters.

The number of coefficients in the cubic symmetry ($\bar{4}3m$) is given by Eq. (4-31) as it applies to the character of Eq. (13-22). Using the table of symmetry classes for ($\bar{4}3m$) in Appendix 5, we obtain the tabulation

Symmetry elements	1	$3[2^1]$	$6[\bar{4}]$	$6[m^-]$	$8[3]$
ϕ	0	π	$\frac{1}{2}\pi$	π	$\frac{2}{3}\pi$
$\chi(\phi)$	56	8	0	8	2

and Eq. (4-31) becomes

$$n_{(\bar{4}3m)} = \frac{1}{24}[1\cdot 1\cdot 56 + 3\cdot 1\cdot 8 + 6\cdot 1\cdot 0 + 6\cdot 1\cdot 8 + 8\cdot 1\cdot 2] = 6$$

Hence there are six independent coefficients characterizing a cubic crystal, whereas there are only three such coefficients in isotropic material.

The explicit scheme of coefficients in ($\bar{4}3m$) is obtained by applying the direct inspection method of Chapter 3. From Appendix 2 the generating elements of ($\bar{4}3m$) are seen to be B and $\bar{4}$. The threefold rotation B around the body diagonal of a cube establishes the equivalence of the three coordinate axes. The fourfold rotation 4 (a sixth rank polar tensor does not distinguish between 4 and $\bar{4}$) eliminates a large number of coefficients. This reduction is carried out in Table 13-1 in two stages. Its first three columns list the total of 56 coefficients, of which 20 are independent. The last column gives the final reduction leading to the set of 20 nonvanishing entries involving 6 constants

$$(111) = (222) = (333)$$

$$(112) = (113) = (122) = (133) = (223) = (233)$$

$$(123)$$

$$(144) = (255) = (366)$$

$$(155) = (166) = (244) = (266) = (344) = (355)$$

$$(456)$$

(13-24)

13.5 Formulation of Higher-Order Interactions

Table 13-1.
Reduction of the Third-Order Elastic Coefficient c_{ijk} in the Symmetry Group ($\bar{4}3m$) by the Direct Inspection Method

Tensor coefficient	(B) $1 \to 2 \to 3 \to 1$ $4 \to 5 \to 6 \to 4$			(4) $1 \to 2,\ 2 \to 1,\ 3 \to 3$ $4 \to -5,\ 5 \to 4,\ 6 \to -6$
111	=	222	= 333	
444		555	666 =	0
112		223	133	113
113		122	233	112
114		225	336	0
115		226	334	0
116		224	335	0
123		123		
124		235	136	0
125		236	134	0
126		234	135	0
144		255	366	
145		256	346	0
146		245	356	0
155		266	344	166
156		246	345	0
166		244	355	155
445		556	466	0
455		566	446	0
456		456		

In isotropic materials, the scheme of constants given in Eq. (13-24) must be further reduced by three relations between the independent coefficients. These relations follow by imposing the additional rotational symmetry of 45° around the z axis. Under this rotation the indices of c_{ijk} transform as follows.

$$(1) \to \frac{1}{2}[(1) + 2(6) + (2)] \qquad (4) \to \frac{1}{\sqrt{2}}[-(5) + (4)]$$

$$(2) \to \frac{1}{2}[(1) - 2(6) + (2)] \qquad (5) \to \frac{1}{\sqrt{2}}[(5) + (4)]$$

$$(3) \to (3) \qquad\qquad\qquad\qquad (6) \to \frac{1}{2}[-(1) + (2)]$$

Hence the coefficient (111) becomes $\frac{1}{8}[(1) + 2(6) + (2)]^3$ or

$$(111) = \frac{1}{8}[(111) + 6(116) + 3(112) + 12(166) + 12(126) + 3(122)$$
$$+ 8(666) + 12(266) + 6(226) + (222)]$$

and using the zeros and identities of Table 13-1 and Eq. (13-24) we obtain the relation

$$(111) = (122) + 4(155)$$

Similarly,

$$(123) = \frac{1}{4}\{[(1) + (2)]^2 - 4(66)\}(3)$$

leading to

$$(123) = (122) - 2(144)$$

Finally,

$$(456) = \frac{1}{4}[-(1) + (2)][(44) - (55)]$$

giving the relation

$$2(456) = (155) - (144)$$

If we consider (123), (144), and (456) as the *independent* coefficients in isotropic material, the relations reducing Eq. (13-24) to isotropic material are

$$(111) = (123) + 6(144) + 8(456)$$
$$(122) = (123) + 2(144) \tag{13-25}$$
$$(155) = (144) + 2(456)$$

In order to express the elastic parameters of polycrystalline material in terms of single crystal constants we make use of the invariants of Eq. (13-18). By carrying out the multiplication of vector components indicated in that equation, the invariants become explicitly

13.5 Formulation of Higher-Order Interactions

$$I_1(3) = [(111) + (222) + (333)] + 3[(122) + (133) + (112) + (233) \\ + (113) + (223)] + 6(123)$$

$$I_2(3) = [(111) + (222) + (333)] + 3[(155) + (166) + (244) + (266) \\ + (344) + (355)] + 6(456)$$

$$I_3(3) = [(111) + (222) + (333)] + [(122) + (133) + (112) + (233) \\ + (113) + (223)] + 2[(155) + (166) + (244) + (266) \\ + (344) + (355)] + 2[(144) + (255) + (366)] \quad (13\text{-}26)$$

Equation (13-26) is valid for *all* crystal symmetries. It indicates that of the 56 possible elastic parameters, only 20 make contributions to the averages in a polycrystalline medium. By using Eqs. (13-24) and (13-25) in Eq. (13-26), we can determine the invariants of the isotropic solid. If we label the elastic coefficients of the isotropic medium by capital letters, we obtain, in terms of the three independent coefficients of this medium

$$I_1 = 27C_{123} + 54C_{144} + 24C_{456}$$

$$I_2 = 3C_{123} + 36C_{144} + 66C_{456}$$

$$I_3 = 9C_{123} + 48C_{144} + 48C_{456}$$

If these equations are inverted, we obtain

$$C_{123} = \frac{1}{210}[16I_1 + 16I_2 - 30I_3]$$

$$C_{144} = \frac{1}{210}[-5I_1 - 12I_2 + 19I_3] \quad (13\text{-}27)$$

$$C_{456} = \frac{1}{210}[2I_1 + 9I_2 - 9I_3]$$

Equation (13-27) is a general formulation for the elastic parameters to be expected in a polycrystalline medium under uniform strain. It applies to single crystallites of any symmetry group when the invariants I_1, I_2, I_3 are expressed in terms of the single crystal constants of that group according to Eq. (13-26).

In particular, for the group $(\bar{4}3m)$ these invariants take the form

$$I_1 = 3c_{111} + 18c_{122} + 6c_{123}$$

$$I_2 = 3c_{111} + 18c_{155} + 6c_{456}$$

$$I_3 = 3c_{111} + 6c_{122} + 12c_{155} + 6c_{144}$$

and therefore a polycrystalline aggregate of such cubic crystallites has third-order elastic parameters given by

$$C_{123} = \frac{1}{35}[c_{111} + 18c_{122} - 12c_{155} + 16c_{123} + 16c_{456} - 30c_{144}]$$

$$C_{144} = \frac{1}{35}[c_{111} + 4c_{122} + 2c_{155} - 5c_{123} - 12c_{456} + 19c_{144}] \qquad (13\text{-}28)$$

$$C_{456} = \frac{1}{35}[c_{111} - 3c_{122} + 9c_{155} + 2c_{123} + 9c_{456} - 9c_{144}]$$

where the coefficients on the right-hand side are the constants of the cubic crystal.

As discussed in Chapter 11, the foregoing results hold strictly only if the polycrystalline medium is under uniform strain. In the opposite limiting case of uniform stress, the same calculation can be carried out using the constants s_{ijk}. This involves additional factors of $\frac{1}{2}$ in front of any coefficient, once for every occurrence of an index between 4 and 6.

13.6. POLYCRYSTALLINE AXIALLY SYMMETRIC MEDIA

When the polycrystalline aggregate contains single crystallites that all retain one common direction but have otherwise random orientations about this common axis, we obtain a cylindrically symmetric medium. In practice, this is most likely to occur in crystals that have a strongly preferred axis, which is usually an axis of symmetry in the single crystal. However, it is no more difficult to discuss the formulation for a common direction completely arbitrary with respect to crystal axes. Since it is convenient to tie the coordinate system to this common direction, all tensors will appear in general form. We therefore discuss the problem of triclinic symmetry.

The number of invariants of any tensor expected in cylindrical symmetry has been given by Eq. (13-16). As already mentioned, this number is an upper limit for the number of independent tensor coefficients to be expected in axial polycrystalline media, because the existence of additional symmetry elements in the finite crystal groups reduces the number of independent invariants by making some of them identical.

The invariants in cylindrical symmetry are constructed in the same manner as the rotational invariants of the isotropic solid by determining the independent scalars that can be formed from the products of vectors identified with the tensor indices. The important difference is that here only rotations around the z axis are allowed, so that no transformations couple the x and y components

13.6 Formulation of Higher-Order Interactions

with the z components of the vectors. This restriction results in a larger number of invariants in cylindrical media.

The cylindrical invariants of a second rank tensor have been worked out in Problem 4-4 by breaking up the vectors $\mathbf{V}_1$ and $\mathbf{V}_2$ into components normal and parallel to the axis of rotation. If these are $\mathbf{v}$ and $\mathbf{w}$, respectively, and we consider a symmetrical second rank tensor, the only scalars that can be constructed are the scalar products of each of the subvectors with itself

$$\mathbf{v} \cdot \mathbf{v} = (x^2 + y^2), \quad \mathbf{w} \cdot \mathbf{w} = (z^2) \tag{13-29}$$

Hence the dielectric tensor (ϵ_{ij}) has the two cylindrical invariants

$$I_1^c = \epsilon_{11} + \epsilon_{22}, \quad I_2^c = \epsilon_{33} \tag{13-30}$$

and since cylindrical symmetry requires the dielectric tensor scheme

$$(\epsilon_{ij}^c) = \begin{pmatrix} \epsilon_{11}^c & 0 & 0 \\ 0 & \epsilon_{11}^c & 0 \\ 0 & 0 & \epsilon_{33}^c \end{pmatrix} \tag{13-31}$$

we find the general relations

$$\epsilon_{11}^c = \epsilon_{22}^c = \frac{1}{2}(\epsilon_{11} + \epsilon_{22}), \quad \epsilon_{33}^c = \epsilon_{33} \tag{13-32}$$

where the coefficients on the right-hand side refer to the diagonal single crystal constants expressed in the coordinate system that has the z axis as the axis of cylindrical symmetry.

As a second example we treat the elastic constants tensor of Eq. (13-1). It follows from Eq. (13-16) that the number of cylindrical invariants is 5, and as already discussed in Section 13-3, we expect that the scheme of (c_{ij}^c) is identical with that of the group (6)

$$c_{ij}^c = \begin{pmatrix} c_{11}^c & c_{12}^c & c_{13}^c & 0 & 0 & 0 \\ c_{12}^c & c_{11}^c & c_{13}^c & 0 & 0 & 0 \\ c_{13}^c & c_{13}^c & c_{33}^c & 0 & 0 & 0 \\ 0 & 0 & 0 & c_{44}^c & 0 & 0 \\ 0 & 0 & 0 & 0 & c_{44}^c & 0 \\ 0 & 0 & 0 & 0 & 0 & \frac{1}{2}(c_{11}^c - c_{12}^c) \end{pmatrix} \quad (13\text{-}33)$$

The cylindrical invariants are the five independent scalars obtained by combining the two vectors $\mathbf{V}_1 = (\mathbf{v}_1, \mathbf{w}_1)$, $\mathbf{V}_2 = (\mathbf{v}_2, \mathbf{w}_2)$, each taken twice

$$I_1 = v_1{}^2 v_2{}^2 = c_{11} + 2c_{12} + c_{22}$$

$$I_2 = w_1{}^2 w_2{}^2 = c_{33}$$

$$I_3 = \frac{1}{2}(v_1{}^2 w_2{}^2 + v_2{}^2 w_1{}^2) = c_{13} + c_{23} \quad (13\text{-}34)$$

$$I_4 = (\mathbf{v}_1 \cdot \mathbf{v}_2)(\mathbf{w}_1 \cdot \mathbf{w}_2) = c_{44} + c_{55}$$

$$I_5 = (\mathbf{v}_1 \times \mathbf{v}_2)^2 = 2c_{12} - 2c_{66}$$

Equation (13-34) establishes the cylindrical invariants for the elastic tensors in all crystal symmetries when the single crystal tensors are expressed in the coordinate system having the z axis along the axial symmetry direction. Only 9 of the 21 elastic parameters contribute to the invariants.

By inserting in the right-hand side of Eq. (13-34) the coefficients of the scheme of Eq. (13-33), we obtain the invariants of the cylindrically symmetric medium. Inversion of this relation expresses the c_{ij}^c in terms of the invariants

$$c_{11}^c = \frac{1}{8}[3I_1 - 2I_5], \quad c_{12}^c = \frac{1}{8}[I_1 + 2I_5]$$

$$c_{33}^c = I_2, \quad c_{13}^c = \frac{1}{2}I_3, \quad c_{44}^c = \frac{1}{2}I_4 \quad (13\text{-}35)$$

These formulas are applied to any specific case by determining the invariants on the right side of Eq. (13-35) in terms of single crystal coefficients according to Eq. (13-34).

For example, if for a group of cubic crystallites the z axis of cubic symmetry becomes the cylindrical axis, the invariants of Eq. (13-34) are

$$I_1 = 2(c_{11} + c_{12}), \quad I_2 = c_{11}, \quad I_3 = 2c_{12}, \quad I_4 = 2c_{44}, \quad I_5 = 2(c_{12} - c_{44})$$

and the cylindrically symmetric combinations of such crystallites have the scheme of elastic parameters given by Eq. (13-33) with the following entries

$$c_{11}^c = \frac{1}{4}(3c_{11} + c_{12} + c_{44}), \quad c_{13}^c = c_{12}$$

$$c_{12}^c = \frac{1}{4}(c_{11} + 3c_{12} - 2c_{44}), \quad c_{33}^c = c_{44}$$

$$\frac{1}{2}(c_{11}^c - c_{12}^c) = \frac{1}{8}(2c_{11} - 2c_{12} + 3c_{44})$$

This scheme has three independent coefficients, but it is not equivalent to either cubic or isotropic symmetry.

All treatments involving simple rotational averages require an assumption concerning the uniformity of strains or stresses. In the foregoing derivation this applies to the strains. If stresses were held uniform, the same method would yield average values of the elastic constants s_{ij}.

Problems

13-1. (a) Show that for the elastic constants (c_{ij}) the additional symmetry relation given by Eq. (13-4) leads to the relations

$$c_{23} = c_{44}, \quad c_{56} = c_{14}$$
$$c_{31} = c_{55}, \quad c_{64} = c_{25}$$
$$c_{12} = c_{66}, \quad c_{45} = c_{36}$$

These are known as the Cauchy relations.

(b) Use the scheme of (c_{ij}) in the crystal symmetry $(2/m)$ to determine the symmetry of the tensor (K_{ijkl}) of Eq. (13-3) in the same crystal.

13-2. (a) Show that the group characters in Table A-4-2 applying to the tensors (c_{ij}) and (P_{ij}) are, respectively,

$$(c_{ij}): \quad 1 - 4\cos^2\phi + 8\cos^3\phi + 16\cos^4\phi$$

$$(P_{ij}): \quad 4\cos^2\phi + 16\cos^3\phi + 16\cos^4\phi$$

(b) Apply the analysis of Section 4-5 to show that in the group (4mm) (c_{ij}) has six and (P_{ij}) has seven independent components.

(c) Show that the additional entry arises because $P_{13} \neq P_{31}$, and that otherwise the schemes (c_{ij}) and (P_{ij}) are identical.

13-3. (a) Show that the group character applying to the third-order Hall effect of Eq. (13-7) is

$$(1 + 2\cos\phi)(-2\cos\phi + 4\cos^2\phi + 8\cos^3\phi)$$

(b) Use the analysis of Section 4-5 to determine that in the group (4mm) there are five independent coefficients $T_{(i)klm}$.

(c) Use the direct inspection method of Chapter 3 to establish the coefficient scheme

	111	112	113	122	123	133	222	223	233	333
(1)	$T_{(1)111}$	0	0	$T_{(1)122}$	0	$T_{(1)133}$	0	0	0	0
(2)	0	0	$T_{(1)122}$	0	0	0	0	$T_{(1)111}$	0	$T_{(1)133}$	0
(3)	0	0	$T_{(3)113}$	0	0	0	0	$T_{3(113)}$	0	$T_{(3)333}$

(d) Show that the same scheme of constants holds for the more symmetric group (6mm) with the additional relation

$$T_{(1)111} = 3T_{(1)122}$$

13-4. In Chapter 10 we developed the modifications of the dielectric constant (ϵ_{ij}) when it becomes linearly dependent on any one of the three vectors $\mathbf{k}$, $\mathbf{E}°$, $\mathbf{B}°$. Extend the treatment to include all possible quadratic terms and determine for each of the tensors in this expansion: (i) The intrinsic symmetry; (ii) the maximum number of independent components; (iii) the optical property it modifies; (iv) the structure of the tensor for cubic symmetry.

13-5. Using any of the forces or fields considered in this book, construct a possible interaction that is represented by the most general fourth rank tensor.

13-6. (a) Discuss the symmetry of the photoelastic tensor of Eq. (13-5) if the material is sensitive not only to elastic stresses but also to torques causing local rotations.

(b) Argue that the same tensor symmetry is obtained for the coefficients of the quadratic dependence on B of the thermoelectric tensor of Chapter 8.

13-7. Formulate the description of piezoconductance as a generalization coupling the phenomena of Chapters 7 and 11, and relate the resulting tensor to other interactions discussed in the text of Chapter 13.

13-8. (a) Show that the group character of the tensor (d_{ijk}) of Eq. (13-8) is given by

$$\chi(\phi) = (1 + 2\cos\phi)(1 - 4\cos^2\phi + 8\cos^3\phi + 16\cos^4\phi)$$

and determine that the number of independent coefficients in the symmetry (4) is 15.

(b) Use direct inspection to determine the tensor in this symmetry and show that it is given by

$$d_{ij1} = \begin{pmatrix} 0 & 0 & 0 & 141 & 151 & 0 \\ & 0 & 0 & 241 & 251 & 0 \\ & & 0 & 341 & 351 & 0 \\ & & & 0 & 0 & 461 \\ & & & & 0 & 561 \\ & & & & & 0 \end{pmatrix}$$

$$d_{ij2} = \begin{pmatrix} 0 & 0 & 0 & 251 & -241 & 0 \\ & 0 & 0 & 151 & -141 & 0 \\ & & 0 & 351 & -341 & 0 \\ & & & 0 & 0 & -561 \\ & & & & 0 & 461 \\ & & & & & 0 \end{pmatrix}$$

$$d_{ij3} = \begin{pmatrix} 113 & 123 & 133 & 0 & 0 & 163 \\ & 113 & 133 & 0 & 0 & -163 \\ & & 333 & 0 & 0 & 0 \\ & & & 443 & 0 & 0 \\ & & & & 443 & 0 \\ & & & & & 663 \end{pmatrix}$$

where only the indices of d_{ijk} are indicated in the entries. Compare this form to the form of the elastic constants tensor (c_{ij}) in the same group.

13-9. If the pyroelectric tensor (δ_i) of Eq. (12-6) depends on even powers of the magnetic field **B**, discuss the symmetry of the tensors describing the quadratic and quartic terms, and relate their symmetry to that of other tensors discussed in the text.

13-10. In analogy with Eq. (12-6) a second-order piezoelectric effect can be written in the phenomenological form

$$D_k = \sum_{i,j} f_{ijk} e_i e_j$$

(a) Determine the intrinsic symmetry of this tensor, and relate it to that of Eq. (13-8).
(b) Show, by writing down the appropriate expression for the free energy density, that (f_{ijk}) and (d_{ijk}) are related.

13-11. (a) Construct the characters of the transformation $\chi(\phi)$ of the three sixth rank tensors R, N, K given by Eqs. (13-11)–(13-13).
(b) Verify from the character the maximum numbers of independent coefficients of each tensor given in the text.
(c) Use the method of Section 4-5 to show that in the symmetry $(\bar{4}3m)$ these tensors have, respectively, 9, 6, and 3 independent coefficients.

13-12. Use the results of Problem 13-11a to determine the number of independent components of the tensors $R, N,$ and K in isotropic media by (i) the irreducible decomposition of their character; (ii) the application of Eq. (13-15).

13-13. (a) Use the results of Problem 13-11a to determine the number of independent components of the tensors $R, N,$ and K in cylindrically symmetric media.
(b) Show that these numbers are two less than the corresponding numbers of coefficients in the group (6).

13-14. (a) Show that the most general tensor of fifth rank has 51 components in an axially symmetric medium.
(b) Verify that this number is also the number of independent coefficients in the group (6).

13-15. (a) Show that in the group (4) the tensor K_{ijklmn} has the nonvanishing components

(111111) = (222222)   (111133) = (222233)
(333333)   (333311) = (333322)
(111112) = −(222221)   (111233) = −(222113)
(111122) = (222211)   (112233)

Formulation of Higher-Order Interactions   189

(b) Show that axial symmetry is induced by requiring invariance under a 45° rotation around the fourfold axis and that it leads to the additional relations

$$(111111) = 5(111122) \quad (111133) = 3(112233)$$
$$(111112) = 0 \quad (111233) = 0$$

13-16. Use the results of Problem 13-12 to construct the invariants for the sixth rank tensors R, N, and K following the formulation given in Eq. (13-18).

13-17. Apply the method of Section 13-5 to determine the fourth-order elastic coefficients c_{ijkl} in cubic, isotropic, and polycrystalline media.
(a) Determine the character of the tensor transformation, and show that in cubic crystals of highest symmetry there are 11 independent coefficients, while in isotropic material there are 4.
(b) Show that the nonvanishing coefficients in cubic and isotropic material are given by the following table.

Cubic	Isotropic
$(1111) = (2222) = (3333)$	$= (1111)$
$(1112) = (2223) = (3331) = (1113) = (1222) = (2333)$	$\frac{1}{4}[(1111) + 3(1122) - 8(4444)]$
$(1144) = (2255) = (3366)$	(1144)
$(1155) = (2266) = (3344) = (2244) = (1166) = (3355)$	(1144)
$(4444) = (5555) = (6666)$	(4444)
$(4455) = (5566) = (6644)$	$\frac{1}{3}(4444)$
$(2233) = (1133) = (1122)$	(1122)
$(3456) = (1564) = (2645)$	0
$(1123) = (2231) = (3312)$	$(1122) - 2(1144)$
$(1244) = (2355) = (3166) = (2155) = (2366) = (3144)$	$(1144) - \frac{2}{3}(4444)$
$(2344) = (1355) = (1266)$	$\frac{1}{4}[(1111) - (1122) - 4(1144)]$

(c) Write out the explicit form of the four invariants of Eq. (13-19) and show that the four independent elastic coefficients of isotropic material can be expressed in terms of the invariants by

$$C_{1111} = \frac{1}{10080}[\,80I_1 - 432I_2 + 96I_3 + 1632I_4\,]$$

$$C_{1122} = \frac{1}{10080}[\,110I_1 + 330I_3 - 36I_3 - 276I_4\,]$$

$$C_{1144} = \frac{1}{10080}[-50I_1 + 186I_2 + 108I_3 - 180I_4\,]$$

$$C_{4444} = \frac{1}{10080} [\, 45I_1 + 135I_2 - 198I_3 + 162I_4 \,]$$

(d) Write the invariants in part (c) for the constants of $(\bar{4}3m)$ and express the four fourth-order elastic parameters of a polycrystalline aggregate of cubic crystallites in terms of the 11 single crystal constants.

13-18. Determine the elastic constants c_{ij}^c of a polycrystalline solid composed of crystallites of cubic symmetry whose (111) axes are parallel, but which are otherwise randomly oriented. Assume uniform strain.

13-19. Determine the elastic constants s_{ij}^c of a polycrystalline solid composed of crystallites of tetragonal symmetry (4) aligned along the fourfold axis.

Bibliography

R. Fieschi and F. G. Fumi, "High-Order Matter Tensors in Symmetrical Systems," *Nuovo Cimento* **10**, 865 (1953).

F. G. Fumi, "Third-Order Elastic Coefficients of Crystals," *Phys. Rev.* **83**, 1274 (1951); *Phys. Rev.* **86**, 561 (1952).

H. Wondratschek, "Tensor Symmetry in the Hexagonal and Rhombohedral Systems, and in Isotropic Media," *N. Jahrbuch Mineral.* **2**, 25 (1953).

W. P. Mason, "Magnetostriction and Anisotropy Energies for Hexagonal, Tetragonal and Orthorhombic Crystals," *Phys. Rev.* **96**, 302 (1954).

H. J. Juretschke, "Third-Order Elastic Constants of Polycrystalline Media," *Appl. Phys. Letters* **12**, 213 (1968).

D. L. Portigal and E. Burstein, "Acoustical Activity and Other First-Order Spatial Dispersion Effects in Crystals," *Phys. Rev.* **170**, 673 (1968).

D. L. Portigal and E. Burstein, "Magnetospatial Dispersion Effects on the Propagation of Electromagnetic Radiation in Crystals," *J. Phys. Chem. Solids* **32**, 603 (1971).

D. F. Nelson and M. Lax, "New Symmetry for Acousto-Optic Scattering," *Phys. Rev. Letters* **24**, 379 (1970).

APPENDIX 1

The 32 Crystallographic Point Groups and Their Symmetry Operations

SYMMETRY ELEMENTS

Table A-1-1[a]

$\dot{2}, \dot{3}, \dot{4}, \dot{6}$	Polar axis of rotation
1, 2, 3, 4, 6	Nonpolar axis of rotation
$m, \bar{1}$	Mirror plane, center of symmetry
$\bar{3}, \bar{4}, \bar{6}$	Axis of rotation–inversion

[a] The notation is that of Hermann and Mauguin.

The relative orientation of an axis or a normal to a mirror plane is indicated by a superscript on the right of the designation for the symmetry element: $^|$ denotes orientation in the direction of a principal axis; $^-$ in the direction of a secondary axis; $^/$ in the direction bisecting two neighboring secondary axes.

Complete details relating to the various combinations of symmetry elements and to the corresponding symmetry operations are given in the references to Chapter 2.

SYMMETRY GROUPS

The abbreviated Hermann–Mauguin notation for the generating elements is enclosed in parentheses:

$$(\bar{1}), \quad (222), \quad (4/mmm), \quad (m3m)$$

For convenience, the Schönflies notation is also given in Table A-1-2.

SYMMETRY OPERATIONS

In Table A-1-2 the different symmetry operations due to a single symmetry element must often be considered separately. The following examples illustrate the notation for symmetry operations.

Given a symmetry element 6^1. The operations generated by this polar sixfold axis are

$$1, \ \dot{6}^1, \ \dot{3}^1, \ [\dot{3}^1]^2, \ [\dot{6}^1]^5$$

If some of the symmetry operations of a single symmetry element fall into the *same class* (i.e., are equivalent under some other symmetry operation, in this case, say, mirror planes m^- and m'), we write

$$1, \ \dot{2}^1, \ 2[\dot{3}^1], \ 2[\dot{6}^1]$$

If two or more symmetry elements are equivalent, the same notation is employed. Thus in the case of three equivalent nonpolar fourfold principal axes we write for the totality of their symmetry operations

$$1, \ 3[2^1], \ 6[4^1]$$

Table A-1-2

Triclinic	(1)	C_1	1
	($\bar{1}$)	C_i	$1, \bar{1}$
Monoclinic	(2)	C_2	$1, \dot{2}$
	(m)	C_s	$1, m$
	(2/m)	C_{2h}	$1, 2, \bar{1}, m^1$
Orthorhombic	(222)	D_2	$1, 2^1, 2^-, 2'$
	(mm)	C_{2v}	$1, \dot{2}^1, m^-, m'$
	(mmm)	D_{2h}	$1, 2^1, 2^-, 2', \bar{1}, m^1, m^-, m'$
Tetragonal	($\bar{4}$)	S_4	$1, 2^1, \bar{4}, [\bar{4}]^3$
	(4)	C_4	$1, \dot{2}^1, \dot{4}^1, [\dot{4}^1]^3$
	(4/m)	C_{4h}	$1, 2^1, 4^1, [4^1]^3, \bar{1}, m^1, \bar{4}, [\bar{4}]^3$
	(42)	D_4	$1, 2^1, 2[4^1], 2[2^-], 2[2']$
	(4mm)	C_{4v}	$1, \dot{2}^1, 2[\dot{4}^1], 2[m^-], 2[m']$
	($\bar{4}2m$)	D_{2d}	$1, 2^1, 2[\bar{4}], 2[2^-], 2[m']$
	(4/mmm)	D_{4h}	$1, 2^1, 2[4^1], 2[2^-], 2[2'], \bar{1}, m^1, 2[\bar{4}],$ $2[m^-], 2[m']$

Table A-1-2 (continued)

Rhombohedral	(3)	C_3	$1, \dot{3}, [\dot{3}]^2$
	$(\bar{3})$	S_6	$1, 3^1, [3^1]^2, \bar{1}, \bar{3}, [\bar{3}]^2$
	$(3m)$	C_{3v}	$1, 2[\dot{3}^1], 3[m^-]$
	(32)	D_3	$1, 2[3^1], 3[\dot{2}^-]$
	$(\bar{3}m)$	D_{3d}	$1, 2[3^1], 3[2^-], \bar{1}, 2[\bar{3}], 3[m^-]$
Hexagonal	$(\bar{6})$	C_{3h}	$1, \bar{6}, 3^1, m^1, [3^1]^2, [\bar{6}]^5$
	(6)	C_6	$1, 6^1, 3^1, 2^1, [3^1]^2, [6^1]^5$
	$(6/m)$	C_{6h}	$1, 6^1, 3^1, 2^1, [3^1]^2, [6^1]^5, \bar{1}, \bar{6}, \bar{3}, m^1, [\bar{3}]^2, [\bar{6}]^5$
	(62)	D_6	$1, 2^1, 2[3^1], 2[6^1], 3[2^-], 3[2']$
	$(6mm)$	C_{6v}	$1, \dot{2}^1, 2[\dot{3}^1], 2[\dot{6}^1], 3[m^-], 3[m']$
	$(\bar{6}2m)$	D_{3h}	$1, m^1, 2[3^1], 2[\bar{6}], 3[2^-], 3[m']$
	$(6/mmm)$	D_{6h}	$1, 2^1, 2[3^1], 2[6^1], 3[2^-], 3[2'], \bar{1}, m^1, 2[\bar{3}], 2[\bar{6}], 3[m^-], 3[m']$
Cubic	(23)	T	$1, 3[2^1], 4[\dot{3}'], 4[\dot{3}']^2$
	$(m3)$	T_h	$1, 3[2^1], 4[3], 4[3]^2, \bar{1}, 3[m^1], 4[\bar{3}], 4[\bar{3}]^2$
	(43)	O	$1, 3[2^1], 6[4^1], 6[2^-], 8[3]$
	$(\bar{4}3m)$	T_d	$1, 3[2^1], 6[\bar{4}], 6[m^-], 8[3]$
	$(m3m)$	O_h	$1, 3[2^1], 6[4^1], 6[2^-], 8[3], \bar{1}, 3[m^1], 6[\bar{4}], 6[m^-], 8[\bar{3}]$

APPENDIX 2

Generating Elements of the 32 Point Groups

Table A-2-1
The Ten Generating Elements and Their Coordinate Transformations[a]

Identity	$1 = \begin{pmatrix} 1 & 0 & 0 \\ 0 & 1 & 0 \\ 0 & 0 & 1 \end{pmatrix}$	Inversion $\bar{1} = \begin{pmatrix} -1 & 0 & 0 \\ 0 & -1 & 0 \\ 0 & 0 & -1 \end{pmatrix}$
Twofold rotation about y axis	$Y = \begin{pmatrix} -1 & 0 & 0 \\ 0 & 1 & 0 \\ 0 & 0 & -1 \end{pmatrix}$	Twofold rotation about z axis $2 = \begin{pmatrix} -1 & 0 & 0 \\ 0 & -1 & 0 \\ 0 & 0 & 1 \end{pmatrix}$
Reflection in xz plane	$m = \begin{pmatrix} 1 & 0 & 0 \\ 0 & -1 & 0 \\ 0 & 0 & 1 \end{pmatrix}$	Reflection in xy plane $M = \begin{pmatrix} 1 & 0 & 0 \\ 0 & 1 & 0 \\ 0 & 0 & -1 \end{pmatrix}$
Threefold rotation about z axis	$3 = \begin{pmatrix} -\frac{1}{2} & \frac{\sqrt{3}}{2} & 0 \\ -\frac{\sqrt{3}}{2} & -\frac{1}{2} & 0 \\ 0 & 0 & 1 \end{pmatrix}$	Fourfold rotation about z axis $4 = \begin{pmatrix} 0 & 1 & 0 \\ -1 & 0 & 0 \\ 0 & 0 & 1 \end{pmatrix}$

Fourfold inversion-rotation about z

$$\bar{4} = \begin{pmatrix} 0 & -1 & 0 \\ 1 & 0 & 0 \\ 0 & 0 & -1 \end{pmatrix}$$

Threefold rotation about cube body diagonal

$$B = \begin{pmatrix} 0 & 1 & 0 \\ 0 & 0 & 1 \\ 1 & 0 & 0 \end{pmatrix}$$

[a] After H. Wondratschek, N. *Jahrbuch Mineral. Monatsh.* **1952**, 217–234.

Table A-2-2
The Generating Elements of the Point Groups[a]

Triclinic	(1)	($\bar{1}$)					
	1	$\bar{1}$					
Monoclinic	(2)	(m)	(2/m)				
	Y	m	$\bar{1}$, Y				
Orthorhombic	(222)	(mm)	(mmm)				
	2, Y	2, m	2, $\bar{1}$, Y				
Rhombohedral	(3)	($\bar{3}$)	(32)	(3m)	($\bar{3}m$)		
	3	3, $\bar{1}$	3, Y	3, m	3, $\bar{1}$, Y		
Tetragonal	($\bar{4}$)	(4)	(4/m)	(42)	(4mm)	(4/mmm)	($\bar{4}2m$)
	$\bar{4}$	4	4, $\bar{1}$	4, Y	4, m	4, $\bar{1}$, Y	$\bar{4}$, Y
Hexagonal	($\bar{6}$)	(6)	(6/m)	(62)	(6mm)	(6/mmm)	($\bar{6}2m$)
	3, M	3, 2	3, 2, $\bar{1}$	3, 2, Y	3, 2, m	3, 2, $\bar{1}$, Y	3, M, Y
Cubic	(23)	(m3)	(43)	($\bar{4}3m$)	(m3m)		
	2, B	2, B, $\bar{1}$	4, B	$\bar{4}$, B	4, B, $\bar{1}$		

[a] After H. Wondratschek, N. *Jahrbuch Mineral. Monatsh.* **1952**, 217–234.

APPENDIX 3

Linear Combinations of Tensor Components Transforming in Invariant Subspaces Under the Covering Operations of the Rotation Group in Three Dimensions

The components are orthogonal, but not normalized.

Table A-3-1
Second Rank Tensor T_{ij} $(i, j = 1,2,3)$

χ_1: $T_{11} + T_{22} + T_{33}$ χ_3: $T_{23} - T_{32}$ χ_5: $T_{11} + T_{22} - 2T_{33}$
$T_{31} - T_{13}$ $T_{11} - T_{22}$
$T_{12} - T_{21}$ $T_{23} + T_{32}$
$T_{31} + T_{13}$
$T_{12} + T_{21}$

Table A-3-2
Third Rank Tensor T_{ij} $(i = 1, \ldots 9; j = 1,2,3)^a$

χ_1: $(T_{41} - T_{71}) + (T_{52} - T_{82}) + (T_{63} - T_{93})$

χ_3: $(T_{53} - T_{83}) - (T_{62} - T_{92})$
$(T_{61} - T_{91}) - (T_{43} - T_{73})$
$(T_{42} - T_{72}) - (T_{51} - T_{81})$

χ_5: $(T_{41} - T_{71}) + (T_{52} - T_{82}) - 2(T_{63} - T_{93})$
$(T_{41} - T_{71}) - (T_{52} - T_{82})$
$(T_{53} - T_{83}) + (T_{62} - T_{92})$
$(T_{61} - T_{91}) + (T_{43} - T_{73})$
$(T_{42} - T_{72}) + (T_{51} - T_{81})$

χ_3: $(T_{53} + T_{83}) + (T_{62} + T_{92}) - 2(T_{21} + T_{31})$
$(T_{61} + T_{91}) + (T_{43} + T_{73}) - 2(T_{12} + T_{32})$
$(T_{42} + T_{72}) + (T_{51} + T_{81}) - 2(T_{13} + T_{23})$

χ_5: $(T_{41} + T_{71}) + (T_{52} + T_{82}) - 2(T_{63} + T_{93})$
$(T_{41} + T_{71}) - (T_{52} + T_{82})$
$(T_{53} + T_{83}) - (T_{62} + T_{92}) - 2(T_{31} - T_{21})$
$(T_{61} + T_{91}) - (T_{43} + T_{73}) - 2(T_{12} - T_{32})$
$(T_{42} + T_{72}) - (T_{51} + T_{81}) - 2(T_{23} - T_{13})$

χ_3: $(T_{53} + T_{83}) + (T_{62} + T_{92}) + (T_{31} + T_{21}) + 3T_{11}$
$(T_{61} + T_{91}) + (T_{43} + T_{73}) + (T_{12} + T_{32}) + 3T_{22}$
$(T_{42} + T_{72}) + (T_{51} + T_{81}) + (T_{23} + T_{13}) + 3T_{33}$

χ_7: $(T_{41} + T_{71}) + (T_{52} + T_{82}) + (T_{63} + T_{93})$
$(T_{53} + T_{83}) + (T_{62} + T_{92}) + (T_{31} + T_{21}) - 2T_{11}$
$(T_{61} + T_{91}) + (T_{43} + T_{73}) + (T_{12} + T_{32}) - 2T_{22}$
$(T_{42} + T_{72}) + (T_{51} + T_{81}) + (T_{23} + T_{13}) - 2T_{33}$
$(T_{53} + T_{83}) - (T_{62} + T_{92}) + (T_{31} - T_{21})$
$(T_{61} + T_{91}) - (T_{43} + T_{73}) + (T_{12} - T_{32})$
$(T_{42} + T_{72}) - (T_{51} + T_{81}) + (T_{23} - T_{13})$

[a] Assignment of index i to index pairs follows Eq. (4-12).

Table A-3-3
Fourth Rank Tensor T_{ij} ($i,j = 1, ..., 9$): Pair-Symmetric Spaces

χ_1: $3(T_{11}+T_{22}+T_{33}) + (T_{23}+T_{32}) + (T_{31}+T_{13}) + (T_{12}+T_{21})$
$+ (T_{44}+T_{47}+T_{74}+T_{77}) + (T_{55}+T_{58}+T_{85}+T_{88}) + (T_{66}+T_{69}+T_{96}+T_{99})$

χ_5: $6(T_{11}+T_{22}-2T_{33}) - (T_{23}+T_{32}) - (T_{31}+T_{13}) + 2(T_{12}+T_{21})$
$- (T_{44}+T_{47}+T_{74}+T_{77}) - (T_{55}+T_{58}+T_{85}+T_{88}) + 2(T_{66}+T_{69}+T_{96}+T_{99})$

$6(T_{11}-T_{22}) - (T_{23}+T_{32}) + (T_{31}+T_{13})$
$- (T_{44}+T_{47}+T_{74}+T_{77}) + (T_{55}+T_{58}+T_{85}+T_{88})$

$(T_{14}+T_{17}+T_{41}+T_{71}) + 3(T_{24}+T_{27}+T_{42}+T_{72}) + 3(T_{34}+T_{37}+T_{43}+T_{73})$
$+ (T_{56}+T_{59}+T_{65}+T_{95}) + (T_{68}+T_{98}+T_{86}+T_{89})$

$(T_{25}+T_{28}+T_{52}+T_{82}) + 3(T_{35}+T_{38}+T_{53}+T_{83}) + 3(T_{15}+T_{18}+T_{51}+T_{81})$
$+ (T_{67}+T_{64}+T_{76}+T_{46}) + (T_{79}+T_{49}+T_{97}+T_{94})$

$(T_{36}+T_{39}+T_{63}+T_{93}) + 3(T_{16}+T_{19}+T_{61}+T_{91}) + 3(T_{26}+T_{29}+T_{62}+T_{92})$
$+ (T_{78}+T_{75}+T_{87}+T_{57}) + (T_{84}+T_{54}+T_{48}+T_{45})$

χ_9: $2(T_{11}+T_{22}+T_{33}) - (T_{23}+T_{32}) - (T_{31}+T_{13}) - (T_{12}+T_{21})$
$- (T_{44}+T_{47}+T_{74}+T_{77}) - (T_{55}+T_{58}+T_{85}+T_{88}) - (T_{66}+T_{69}+T_{96}+T_{99})$

$(T_{11}+T_{22}-2T_{33}) + (T_{23}+T_{32}) + (T_{31}+T_{13}) - 2(T_{12}+T_{21})$
$+ (T_{44}+T_{47}+T_{74}+T_{77}) + (T_{55}+T_{58}+T_{85}+T_{88}) - 2(T_{66}+T_{69}+T_{96}+T_{99})$

$(T_{11}-T_{22}) + (T_{23}+T_{32}) - (T_{31}+T_{13})$
$+ (T_{44}+T_{47}+T_{74}+T_{77}) - (T_{55}+T_{58}+T_{85}+T_{88})$

$2(T_{14}+T_{17}+T_{41}+T_{71}) - (T_{24}+T_{27}+T_{42}+T_{72}) - (T_{34}+T_{37}+T_{43}+T_{73})$
$+ 2(T_{56}+T_{59}+T_{65}+T_{95}) + 2(T_{68}+T_{98}+T_{86}+T_{89})$

$2(T_{25}+T_{28}+T_{52}+T_{82}) - (T_{35}+T_{38}+T_{53}+T_{83}) - (T_{15}+T_{18}+T_{51}+T_{81})$
$+ 2(T_{67}+T_{64}+T_{76}+T_{46}) + 2(T_{79}+T_{49}+T_{97}+T_{94})$

$2(T_{36}+T_{39}+T_{63}+T_{93}) - (T_{16}+T_{19}+T_{61}+T_{91}) - (T_{26}+T_{29}+T_{62}+T_{92})$
$+ 2(T_{78}+T_{75}+T_{87}+T_{57}) + 2(T_{84}+T_{54}+T_{48}+T_{45})$

$(T_{24}+T_{27}+T_{42}+T_{72}) - (T_{34}+T_{37}+T_{43}+T_{73})$

$(T_{35}+T_{38}+T_{53}+T_{83}) - (T_{15}+T_{18}+T_{51}+T_{81})$

$(T_{16}+T_{19}+T_{61}+T_{91}) - (T_{26}+T_{29}+T_{62}+T_{92})$

χ_1: $2(T_{23}+T_{32}) + 2(T_{31}+T_{13}) + 2(T_{12}+T_{21}) - (T_{44}+T_{47}+T_{74}+T_{77})$
$- (T_{55}+T_{58}+T_{85}+T_{88}) - (T_{66}+T_{69}+T_{96}+T_{99})$

χ_5: $-2(T_{23}+T_{32}) - 2(T_{31}+T_{13}) + 4(T_{12}+T_{21}) + (T_{44}+T_{47}+T_{74}+T_{77})$
$+ (T_{55}+T_{58}+T_{85}+T_{88}) - 2(T_{66}+T_{69}+T_{96}+T_{99})$

$-2(T_{23}+T_{32}) + 2(T_{31}+T_{13}) + \qquad\qquad (T_{44}+T_{47}+T_{74}+T_{77})$
$- (T_{55}+T_{58}+T_{85}+T_{88})$

$2(T_{14}+T_{17}+T_{41}+T_{71}) - (T_{56}+T_{59}+T_{65}+T_{95}) - (T_{68}+T_{98}+T_{86}+T_{89})$

$2(T_{25}+T_{28}+T_{52}+T_{82}) - (T_{67}+T_{64}+T_{76}+T_{46}) - (T_{79}+T_{49}+T_{97}+T_{94})$

$2(T_{36}+T_{39}+T_{63}+T_{93}) - (T_{78}+T_{75}+T_{87}+T_{57}) - (T_{84}+T_{54}+T_{48}+T_{45})$

χ_3: $(T_{24}+T_{27}-T_{42}-T_{72}) - (T_{34}+T_{37}-T_{43}-T_{73}) - (T_{56}+T_{59}-T_{65}-T_{95})$
$+ (T_{68}+T_{98}-T_{86}-T_{89})$

$(T_{35}+T_{38}-T_{53}-T_{83}) - (T_{15}+T_{18}-T_{51}-T_{81}) - (T_{67}+T_{64}-T_{76}-T_{46})$
$+ (T_{79}+T_{49}-T_{97}-T_{94})$

$(T_{16}+T_{19}-T_{61}-T_{91}) - (T_{26}+T_{29}-T_{62}-T_{92}) - (T_{78}+T_{75}-T_{87}-T_{57})$
$+ (T_{84}+T_{54}-T_{48}-T_{45})$

χ_5: $(T_{23}-T_{32}) + (T_{31}-T_{13}) - 2(T_{12}-T_{21})$

$(T_{23}-T_{32}) - (T_{31}-T_{13})$

$(T_{14}+T_{17}-T_{41}-T_{71}) + (T_{24}+T_{27}-T_{42}-T_{72}) + (T_{34}+T_{37}-T_{43}-T_{73})$

$(T_{25}+T_{28}-T_{52}-T_{82})+(T_{35}+T_{38}-T_{53}-T_{83})+(T_{15}+T_{18}-T_{51}-T_{81})$

$(T_{36}+T_{39}-T_{63}-T_{93})+(T_{16}+T_{19}-T_{61}-T_{91})+(T_{26}+T_{29}-T_{62}-T_{92})$

χ_7: $(T_{23}-T_{32})+(T_{31}-T_{13})+(T_{12}-T_{21})$

$(T_{24}+T_{27}-T_{42}-T_{72})-(T_{34}+T_{37}-T_{43}-T_{73})+(T_{56}+T_{59}-T_{65}-T_{95})$
$-(T_{68}+T_{98}-T_{86}-T_{89})$

$(T_{35}+T_{38}-T_{53}-T_{83})-(T_{15}+T_{18}-T_{51}-T_{81})+(T_{67}+T_{64}-T_{76}-T_{46})$
$-(T_{79}+T_{49}-T_{97}-T_{94})$

$(T_{16}+T_{19}-T_{61}-T_{91})-(T_{26}+T_{29}-T_{62}-T_{92})+(T_{78}+T_{75}-T_{87}-T_{57})$
$-(T_{84}+T_{54}-T_{48}-T_{45})$

$2(T_{14}+T_{17}-T_{41}-T_{71})-(T_{24}+T_{27}-T_{42}-T_{74})-(T_{34}+T_{37}-T_{43}-T_{73})$

$2(T_{25}+T_{28}-T_{52}-T_{82})-(T_{35}+T_{38}-T_{53}-T_{83})-(T_{15}+T_{18}-T_{51}-T_{81})$

$2(T_{36}+T_{39}-T_{63}-T_{93})-(T_{16}+T_{19}-T_{61}-T_{91})-(T_{26}+T_{29}-T_{62}-T_{92})$

APPENDIX 4

Characters of Representations of the Group of Rotations in Three Dimensions

The following are tables of the characters of various representations of the rotation group in three dimensions discussed in Section 4-4.

Table A-4-1
Characters of the Irreducible Representations

$\chi_1(\phi) = 1$
$\chi_3(\phi) = 1 + 2\cos\phi$
$\chi_5(\phi) = -1 + 2\cos\phi + 4\cos^2\phi$
$\chi_7(\phi) = -1 - 4\cos\phi + 4\cos^2\phi + 8\cos^3\phi$
$\chi_9(\phi) = 1 - 4\cos\phi - 12\cos^2\phi + 8\cos^3\phi + 16\cos^4\phi$
$\chi_{11}(\phi) = 1 + 6\cos\phi - 12\cos^2\phi - 32\cos^3\phi + 16\cos^4\phi + 32\cos^5\phi$
$\chi_{13}(\phi) = -1 + 6\cos\phi + 24\cos^2\phi - 32\cos^3\phi - 80\cos^4\phi + 32\cos^5\phi + 64\cos^6\phi$
$\chi_{15}(\phi) = -1 - 8\cos\phi + 24\cos^2\phi + 80\cos^3\phi - 80\cos^4\phi - 192\cos^5\phi + 64\cos^6\phi + 128\cos^7\phi$
$\chi_{17}(\phi) = 1 - 8\cos\phi - 40\cos^2\phi + 80\cos^3\phi + 240\cos^4\phi - 192\cos^5\phi - 448\cos^6\phi + 128\cos^7\phi + 256\cos^8\phi$

Appendix 4

Table A-4-2
Class Characters for Representations of the Rotation Group in Three Dimensions Based on the Vectorial Representation[a]

$$\chi_3(\phi) = 1 + 2\cos\phi$$
$$\chi_{(3\times 3)}(\phi) = 1 + 4\cos\phi + 4\cos^2\phi$$
$$\chi_{(3\times 3)_s}(\phi) = 2\cos\phi + 4\cos^2\phi$$
$$\chi_{(3\times 3\times 3)}(\phi) = 1 + 6\cos\phi + 12\cos^2\phi + 8\cos^3\phi$$
$$\chi_{(3\times 3)_s\times 3}(\phi) = 2\cos\phi + 8\cos^2\phi + 8\cos^3\phi$$
$$\chi_{(3\times 3\times 3)_s}(\phi) = -2\cos\phi + 4\cos^2\phi + 8\cos^3\phi$$
$$\chi_{(3\times 3\times 3\times 3)}(\phi) = 1 + 8\cos\phi + 24\cos^2\phi + 32\cos^3\phi + 16\cos^4\phi$$
$$\chi_{(3\times 3)_s\times(3\times 3)_s}(\phi) = 4\cos^2\phi + 16\cos^3\phi + 16\cos^4\phi$$
$$\chi_{((3\times 3)_s\times(3\times 3)_s)_s}(\phi) = 1 - 4\cos^2\phi + 8\cos^3\phi + 16\cos^4\phi$$
$$\chi_{(3\times 3\times 3\times 3)_s}(\phi) = 1 - 2\cos\phi - 8\cos^2\phi + 8\cos^3\phi + 16\cos^4\phi$$
$$\chi_{3^5}(\phi) = 1 + 10\cos\phi + 40\cos^2\phi + 80\cos^3\phi + 80\cos^4\phi + 32\cos^5\phi$$
$$\chi_{(3^5)_s}(\phi) = 1 + 4\cos\phi + 8\cos^2\phi + 24\cos^3\phi + 16\cos^4\phi + 32\cos^5\phi$$
$$\chi_{(3^2)_s\times(3^4)_s}(\phi) = 2\cos\phi - 24\cos^3\phi - 16\cos^4\phi + 64\cos^5\phi + 64\cos^6\phi$$
$$\chi_{(3^2)_s\times((3^2)_s\times(3^2)_s)_s}(\phi) = 2\cos\phi + 4\cos^2\phi - 8\cos^3\phi + 64\cos^5\phi + 64\cos^6\phi$$
$$\chi_{((3)_s^2\times(3^2)_s\times(3^2)_s)_s}(\phi) = 16\cos^2\phi - 8\cos^3\phi - 48\cos^4\phi + 32\cos^5\phi + 64\cos^6\phi$$
$$\chi_{(3^6)_s}(\phi) = 4\cos\phi + 16\cos^2\phi - 24\cos^3\phi - 64\cos^4\phi + 32\cos^5\phi + 64\cos^6\phi$$

[a] For rotation-inversions $\bar{\phi}$, the characters follow the rule of Eq. (4-29).

Table A-4-3
The Irreducible Decomposition of the Characters of Table A-4-2

$$\chi_3(\phi) = \chi_3(\phi)$$
$$\chi_{(3\times 3)}(\phi) = \chi_1(\phi) + \chi_3(\phi) + \chi_5(\phi)$$
$$\chi_{(3\times 3)_s}(\phi) = \chi_1(\phi) + \chi_5(\phi)$$
$$\chi_{(3\times 3\times 3)}(\phi) = \chi_1(\phi) + 3\chi_3(\phi) + 2\chi_5(\phi) + \chi_7(\phi)$$
$$\chi_{(3\times 3)_s\times 3}(\phi) = 2\chi_3(\phi) + \chi_5(\phi) + \chi_7(\phi)$$
$$\chi_{(3\times 3\times 3)_s}(\phi) = \chi_3(\phi) + \chi_7(\phi)$$
$$\chi_{(3\times 3\times 3\times 3)}(\phi) = 3\chi_1(\phi) + 6\chi_3(\phi) + 6\chi_5(\phi) + 3\chi_7(\phi) + \chi_9(\phi)$$
$$\chi_{(3\times 3)_s\times(3\times 3)_s}(\phi) = 2\chi_1(\phi) + \chi_3(\phi) + 3\chi_5(\phi) + \chi_7(\phi) + \chi_9(\phi)$$
$$\chi_{((3\times 3)_s\times(3\times 3)_s)_s}(\phi) = 2\chi_1(\phi) + 2\chi_5(\phi) + \chi_9(\phi)$$
$$\chi_{(3\times 3\times 3\times 3)_s}(\phi) = \chi_1(\phi) + \chi_5(\phi) + \chi_9(\phi)$$
$$\chi_{3^5}(\phi) = 6\chi_1(\phi) + 15\chi_3(\phi) + 15\chi_5(\phi) + 10\chi_7(\phi) + 4\chi_9(\phi) + \chi_{11}(\phi)$$
$$\chi_{(3^5)_s}(\phi) = \chi_3(\phi) + \chi_7(\phi) + \chi_{11}(\phi)$$
$$\chi_{(3^2)_s\times(3^4)_s}(\phi) = 2\chi_1(\phi) + \chi_3(\phi) + 4\chi_5(\phi) + 2\chi_7(\phi) + 3\chi_9(\phi) + \chi_{11}(\phi) + \chi_{13}(\phi)$$
$$\chi_{(3^2)_s\times((3^2)_s\times(3^2)_s)_s}(\phi) = 4\chi_1(\phi) + 2\chi_3(\phi) + 7\chi_5(\phi) + 3\chi_7(\phi) + 4\chi_9(\phi) + \chi_{11}(\phi) + \chi_{13}(\phi)$$
$$\chi_{((3^2)_s\times(3^2)_s\times(3^2)_s)_s}(\phi) = 3\chi_1(\phi) + 3\chi_5(\phi) + \chi_7(\phi) + 2\chi_9(\phi) + \chi_{13}(\phi)$$
$$\chi_{(3^6)_s}(\phi) = \chi_1(\phi) + \chi_5(\phi) + \chi_9(\phi) + \chi_{13}(\phi)$$

Appendix 4

Table A-4-4
Formulas for Characters of Multiple Totally Symmetric
Product Representations

Double	$\chi_{(l^2)_s}(\phi) = \frac{1}{2}\chi_l(2\phi) + \frac{1}{2}\chi_l^2(\phi)$
Triple	$\chi_{(l^3)_s}(\phi) = \frac{1}{3}\chi_l(3\phi) + \frac{1}{2}\chi_l(2\phi)\chi_l(\phi) + \frac{1}{6}\chi_l^3(\phi)$
Quadruple	$\chi_{(l^4)_s}(\phi) = \frac{1}{4}\chi_l(4\phi) + \frac{1}{3}\chi_l(3\phi)\chi_l(\phi) + \frac{1}{8}\chi_l^2(2\phi)$ $+ \frac{1}{4}\chi_l(2\phi)\chi_l^2(\phi) + \frac{1}{24}\chi_l^4(\phi)$
Quintuple	$\chi_{(l^5)_s}(\phi) = \frac{1}{5}\chi_l(5\phi) + \frac{1}{4}\chi_l(4\phi)\chi_l(\phi) + \frac{1}{6}\chi_l(3\phi)\chi_l^2(\phi)$ $+ \frac{1}{6}\chi_l(3\phi)\chi_l(2\phi) + \frac{1}{12}\chi_l(2\phi)\chi_l^3(\phi)$ $+ \frac{1}{8}\chi_l^2(2\phi)\chi_l(\phi) + \frac{1}{120}\chi_l^5(\phi)$
Sixtuple	$\chi_{(l^6)_s}(\phi) = \frac{1}{6}\chi_l(6\phi) + \frac{1}{5}\chi_l(5\phi)\chi_l(\phi) + \frac{1}{8}\chi_l(4\phi)\chi_l^2(\phi)$ $+ \frac{1}{8}\chi(4\phi)\chi_l(2\phi) + \frac{1}{18}\chi_l^2(3\phi)$ $+ \frac{1}{6}\chi_l(3\phi)\chi_l(2\phi)\chi_l(\phi) + \frac{1}{18}\chi_l(3\phi)\chi_l^3(\phi)$ $+ \frac{1}{48}\chi_l^3(2\phi) + \frac{1}{16}\chi_l(2\phi)\chi_l^2(\phi)$ $+ \frac{1}{48}\chi_l(2\phi)\chi_l^4(\phi) + \frac{1}{720}\chi_l^6(\phi)$

Table A-4-5
Characters of Totally Antisymmetric Product Representations[a]

$$\chi_{(l \times l)_{\text{antis}}}(\phi) = \frac{1}{2}\chi_l^{\,2}(\phi) - \frac{1}{2}\chi_l(2\phi)$$

$$\chi_{(l \times l \times l)_{\text{antis}}}(\phi) = \frac{1}{6}\chi_l^{\,3}(\phi) - \frac{1}{2}\chi_l(\phi)\chi_l(2\phi) + \frac{1}{3}\chi_l(3\phi)$$

[a] $(l \times l)_{\text{antis}}$ and $(l \times l \times l)_{\text{antis}}$ are to be interpreted as antisymmetric index pairs $(ij) = -(ji)$ and triplets $(ijk) = -(jik) = -(ikj)$, respectively, occurring within a tensor. For $l = 3$, the foregoing antisymmetric tensors become an axial vector and a pseudoscalar.

APPENDIX 5

Irreducible Characters and Character Tables of the 32 Crystallographic Point Groups

Entries in the following tables are arranged according to crystal systems. The crystal symmetry group is indicated at the top of each table. All the crystal operations, arranged in classes, are given on the right. The number preceding each symmetry operation is the number of equivalent operations forming the given class.

In the body of each table, the class characters of all the irreducible representations of the group appear at the right. At the left are letter symbols which follow the scheme: A, one-dimensional representation symmetric for all rotations around a principal axis; B, one-dimensional representation antisymmetric for all rotations around such axis; E, two-dimensional representation (or two complex conjugate one-dimensional representations); T, three-dimensional representation.

The first footnote to each table indicates another point group not included in any table. Each of these groups contains the same symmetry operations and classes as some group in the table, plus an equal number of others generated by introducing a center of symmetry $\bar{1}$ into the original symmetry. Each point group $G \times \bar{1}$ has twice the number of classes of G, and correspondingly two irreducible representations for any one of G. For example, corresponding to A_1 and A_2 of (62) we have for (6/mmm) the representations A_{1g}, A_{1u}, A_{2g}, A_{2u}. Representations A_{1g} and A_{1u} have the same characters as A_1 for the classes common to (62) and (6/mmm), while the characters of A_{1g} and A_{1u} for the extra classes in (6/mmm) are, respectively, +1 and −1 times the characters of A, for the corresponding classes in (62).

205

Table A-5-1
Triclinic[a]

(1)	1
A	1

[a] Also $(\bar{1}) = (1) \times \bar{1}$.

Table A-5-2
Monoclinic[a]

(2)		1	$\dot{2}$
	(m)	1	m
A	A'	1	1
B	A''	1	−1

[a] Also $(2/m) = (2) \times \bar{1}$.

Table A-5-3
Orthorhombic[a]

(222)		1	$\dot{2}^{|}$	2^-	$2^{/}$
	(mm)	1	$\dot{2}^{|}$	m^-	$m^{/}$
A_1	A_1	1	1	1	1
B_1	A_2	1	1	−1	−1
B_2	B_1	1	−1	1	−1
B_3	B_2	1	−1	−1	1

[a] Also $(mmm) = (222) \times \bar{1}$.

Table A-5-4
I. Tetragonal[a]

(4)	1	$\dot{2}^{|}$	$\dot{4}^{|}$	$[\dot{4}^{|}]^3$
$(\bar{4})$	1	$2^{|}$	$\bar{4}$	$[\bar{4}]^3$
A	1	1	1	1
B	1	1	−1	−1
E	1	−1	−i	i
	1	−1	i	−i

[a] [a] Also $(4/m) = (4) \times \bar{1}$.

II. Tetragonal[a]

	1	2^I	$2[4^I]$	$2[2^-]$	$2[2']$
(42)					
(4mm)	1	$\dot{2}^I$	$2[\dot{4}^I]$	$2[m^-]$	$2[m']$
($\bar{4}2m$)	1	2^I	$2[\bar{4}]$	$2[2^-]$	$2[m']$
A_1	1	1	1	1	1
A_2	1	1	1	-1	-1
B_1	1	1	-1	1	-1
B_2	1	1	-1	-1	1
E	2	-2	0	0	0

[a] Also $(4/mmm) = (42) \times \bar{1}$.

Table A-5-5
I. Rhombohedral[a,b]

(3)	1	$\dot{3}$	$[\dot{3}]^2$
A	1	1	1
E	1	ω	ω^2
	1	ω^2	ω

[a] Also $(\bar{3}) = (3) \times \bar{1}$.
[b] Note: $\omega = exp\,(2\pi i/3)^4$.

II. Rhombohedral[a]

	1	$2[3^I]$	$3[\dot{2}^-]$
(32)			
(3m)	1	$2[\dot{3}^I]$	$3[m^-]$
A_1	1	1	1
A_2	1	1	-1
E	2	-1	0

[a] Also $(\bar{3}m) = (32) \times \bar{1}$.

Table A-5-6
I. Hexagonal[a,b]

(6)		1	$\dot{6}^1$	$\dot{3}^1$	$\dot{2}^1$	$[\dot{3}^1]^2$	$[\dot{6}^1]^5$
	$(\bar{6})$	1	$\bar{6}$	3^1	m^1	$[3^1]^2$	$[\bar{6}]^5$
A	A'	1	1	1	1	1	1
B	A''	1	-1	1	-1	1	-1
E*	E'	1	ω^2	$-\omega$	1	ω^2	$-\omega$
		1	$-\omega$	ω^2	1	$-\omega$	ω^2
E**	E''	1	ω	ω^2	-1	$-\omega$	$-\omega^2$
		1	$-\omega^2$	$-\omega$	-1	ω^2	ω

[a] Also $(6/m) = (6) \times \bar{1}$.
[b] Note: $\omega = \exp(2\pi i/6) = -\omega^4$.

II. Hexagonal[a]

(62)			1	2^1	$2[3^1]$	$2[6^1]$	$3[2^-]$	$3[2']$
	(6mm)		1	$\dot{2}^1$	$2[\dot{3}^1]$	$2[\dot{6}^1]$	$3[m^-]$	$3[m']$
		(62m)	1	m^1	$2[3^1]$	$2[\bar{6}]$	$3[2^-]$	$3[m']$
A_1	A_1	A_1'	1	1	1	1	1	1
A_2	A_2	A_2'	1	1	1	1	-1	-1
B_1	B_2	A_1''	1	-1	1	-1	1	-1
B_2	B_1	A_2''	1	-1	1	-1	-1	1
E*	E*	E'	2	2	-1	-1	0	0
E**	E**	E''	2	-2	-1	1	0	0

[a] Also $(6/mmm) = (62) \times \bar{1}$.

Table A-5-7
I. Cubic[a]

(23)	1	$3[2^1]$	$4[\dot{3}']$	$4[\dot{3}']^2$
A	1	1	1	1
E	1 1	1 1	ω ω^2	ω^2 ω
T	3	-1	0	0

[a] Also $(m3) = (23) \times \bar{1}$.

II. Cubic[b]

(43)	1	$3[2^1]$	$6[4^1]$	$6[2^-]$	$8[3]$
$(\bar{4}3m)$	1	$3[2^1]$	$6[\bar{4}]$	$6[m^-]$	$8[\dot{3}]$
A_1	1	1	1	1	1
A_2	1	1	-1	-1	1
E	2	2	0	0	-1
T_1	3	-1	1	-1	0
T_2	3	-1	-1	1	0

[a] Also $(m3m) = (43) \times \bar{1}$.

APPENDIX 6

The Magnetic Point Groups and Their Symmetry Elements

The magnetic groups are listed under the conventional point groups of Appendix 1 from which they derive. Time-inverted elements are usually denoted by a bar under the element. To simplify notation, we indicate by a plus sign the elements that remain unchanged, and by a minus sign the elements combined with time reversal.

Table A-6-1

Group	Elements	Group	Elements
(1)	1	(mm)	$1, 2^I, m, m'$
$(\bar{1})$	$1, \bar{1}$	$(m\underline{m})$	$+, -, +, -$
$(\underline{\bar{1}})$	$+, -$	$(\underline{mm})$	$+, +, -, -$
(2)	$1, 2$	(mmm)	$1, 2^I, 2^-, 2', \bar{1}, m^I, m^-, m'$
$(\underline{2})$	$+, -$	$(mm\underline{m})$	$+, -, -, +, -, +, +, -$
(m)	$1, m$	$(m\underline{mm})$	$+, +, -, -, +, +, -, -$
$(\underline{m})$	$+, -$	$(\underline{mmm})$	$+, +, +, +, -, -, -, -$
$(2/m)$	$1, 2, \bar{1}, m^I$	$(\bar{4})$	$1, 2^I, \bar{4}, (\bar{4})^3$
$(2/\underline{m})$	$+, +, -, -$	$(\underline{\bar{4}})$	$+, +, -, -$
$(\underline{2}/m)$	$+, -, -, +$	(4)	$1, 2^I, 4^I, (4^I)^3$
$(\underline{2}/\underline{m})$	$+, -, +, -$	$(\underline{4})$	$+, +, -, -$
(222)	$1, 2^I, 2^-, 2'$		
$(2\underline{22})$	$+, +, -, -$		

Appendix 6

Table A-6-1 (continued)

Group	Elements
$(4/m)$	$1,\ 2^1,\ 4^1,\ (4^1)^3,\ \bar{1},\ m^1,\ \bar{4},\ (\bar{4})^3$
$(4/\underline{m})$	+, +, +, +, −, −, −, −
$(\underline{4}/m)$	+, +, −, −, +, +, −, −
$(\underline{4}/\underline{m})$	+, +, −, −, −, −, +, +
(42)	$1,\ 2^1,\ 2(4^1),\ 2(2^-),\ 2(2')$
$(4\underline{2})$	+, +, +, −, −
$(\underline{42})$	+, +, −, +, −
$(4mm)$	$1,\ 2^1,\ 2(4^1),\ 2(m^-),\ 2(m')$
$(4\underline{mm})$	+, +, +, −, −
$(\underline{4mm})$	+, +, −, +, −
$(\bar{4}2m)$	$1,\ 2^1,\ 2(\bar{4}),\ 2(2^-),\ 2(m')$
$(\bar{4}\underline{2m})$	+, +, +, −, −
$(\bar{4}\underline{2}m)$	+, +, −, +, −
$(\underline{\bar{4}2m})$	+, +, −, −, +
$(4/mmm)$	$1,\ 2^1,\ 2(4^1),\ 2(2^-),\ 2(2'),\ \bar{1},\ m^1,\ 2(\bar{4}),\ 2(m^-),\ 2(m')$
$(4/\underline{m}mm)$	+, +, +, −, −, −, −, −, +, +
$(4/m\underline{mm})$	+, +, +, −, −, +, +, +, −, −
$(4/\underline{mmm})$	+, +, +, +, +, −, −, −, −, −
$(\underline{4}/m\underline{mm})$	+, +, −, +, −, +, +, −, +, −
$(\underline{4}/\underline{mm}\underline{m})$	+, +, −, −, +, −, −, +, +, −
(3)	$1,\ 3,\ (3)^2$
$(\bar{3})$	$1,\ 3^1,\ (3^1)^2,\ \bar{1},\ \bar{3},\ (\bar{3})^2$
$(\underline{\bar{3}})$	+, +, +, −, −, −
$(3m)$	$1,\ 2(3^1),\ 3(m^-)$
$(3\underline{m})$	+, +, −
(32)	$1,\ 2(3^1),\ 3(2^-)$
$(3\underline{2})$	+, +, −
$(\bar{3}m)$	$1,\ 2(3^1),\ 3(2^-),\ \bar{1},\ 2(\bar{3}),\ 3(m^-)$
$(\bar{3}\underline{m})$	+, +, −, +, +, −
$(\underline{\bar{3}}m)$	+, +, −, −, −, +
$(\underline{\bar{3}}\underline{m})$	+, +, +, −, −, −

Table A-6-1 (continued)

Group	Elements
$(\bar{6})$	$1, \bar{6}, 3^1, m^1, (3^1)^2, (\bar{6})^5$
$(\underline{\bar{6}})$	$+, -, +, -, +, \quad -$
$(\dot{6})$	$1, \dot{6}^1, \dot{3}^1, \dot{2}^1, (\dot{3}^1)^2, (\dot{6}^1)^5$
$(\underline{\dot{6}})$	$+, -, +, -, +, \quad -$
$(6/m)$	$1, 6^1, 3^1, 2^1, (3^1)^2, (6^1)^5, \bar{1}, \bar{6}, \bar{3}, m^1, (\bar{3})^2, (\bar{6})^5$
$(6/\underline{m})$	$+,+,+,+,+, \quad +, \quad -,-,-,-,-, \quad -$
$(\underline{6}/m)$	$+,-,+,-,+, \quad -, \quad -,-,+,+,-, \quad +$
$(\underline{6}/\underline{m})$	$+,-,+,-,+, \quad -, \quad +,+,-,-,+, \quad -$
(62)	$1, 2^1, 2(3^1), 2(6^1), 3(2^-), 3(2^\prime)$
$(6\underline{2})$	$+,+,+, \quad +, \quad -, \quad -$
$(\underline{6}\underline{2})$	$+,-,+, \quad -, \quad +, \quad -$
$(6mm)$	$1, \dot{2}^1, 2(\dot{3}^1), 2(\dot{6}^1), 3(m^-), 3(m^\prime)$
$(6\underline{mm})$	$+,+,+, \quad +, \quad -, \quad -$
$(\underline{6}\underline{mm})$	$+,-,+, \quad -, \quad -, \quad +$
$(\bar{6}2m)$	$1, m^1, 2(3^1), 2(\bar{6}), 3(2^-), 3(m^\prime)$
$(\bar{6}\underline{2}m)$	$+,+,+, \quad +, \quad -, \quad -$
$(\bar{6}\underline{2}\underline{m})$	$+,-,+, \quad -, \quad +, \quad -$
$(\underline{\bar{6}}\underline{2}m)$	$+,-,+, \quad -, \quad -, \quad +$
$(6/mmm)$	$1, 2^1, 2(3^1), 2(6^1), 3(2^-), 3(2^\prime), \bar{1}, m^1, 2(\bar{3}), 2(\bar{6}), 3(m^-), 3(m^\prime)$
$(6/\underline{mmm})$	$+,+,+, \quad +, \quad -, \quad -, \quad -,-,-, \quad -, \quad +, \quad +$
$(6/m\underline{mm})$	$+,+,+, \quad +, \quad -, \quad -, \quad +,+,+, \quad +, \quad -, \quad -$
$(6/\underline{m}mm)$	$+,+,+, \quad +, \quad +, \quad +, \quad -,-,-, \quad -, \quad -, \quad -$
$(\underline{6}/\underline{m}mm)$	$+,-,+, \quad -, \quad -, \quad +, \quad -,+,-, \quad +, \quad +, \quad -$
$(\underline{6}/m\underline{m}\underline{m})$	$+,-,+, \quad -, \quad +, \quad -, \quad +,-,+, \quad -, \quad +, \quad -$
(23)	$1, 3(2^1), 4(\dot{3}^\prime), 4(\dot{3}^\prime)^2$
$(m3)$	$1, 3(2^1), 4(3), 4(3)^2, \bar{1}, 3(m^1), 4(\bar{3}), 4(\bar{3})^2$
$(\underline{m}3)$	$+,+, \quad +, \quad +, \quad -,-, \quad -, \quad -$
(43)	$1, 3(2^1), 6(4^1), 6(2^-), 8(3)$
$(\underline{43})$	$+,+, \quad -, \quad -, \quad +$
$(\bar{4}3m)$	$1, 3(2^1), 6(\bar{4}), 6(m^-), 8(3)$
$(\bar{4}3\underline{m})$	$+,+, \quad -, \quad -, \quad +$

Table A-6-1 (continued)

Group	Elements								
$(m3m)$	1,	$3(2^1)$,	$6(4^1)$,	$6(2^-)$,	$8(3)$,	$\bar{1}$,	$3(m^1)$,	$6(\bar{4})$,	$6(m^-)$, $8(\bar{3})$
$(\underline{m}3m)$	+, +,	+,	+,	+,	−, −,		−,	−,	−
$(m3\underline{m})$	+, +,	−,	−,	+,	+, +,		−,	−,	+
$(\underline{m}3\underline{m})$	+, +,	−,	−,	+,	−, −,		+,	+,	−

Selected Literature

The references listed here supplement the Bibliography at the end of each chapter. They offer a small additional sampling of the literature touching on the material of the text, without presenting a systematic or representative coverage. Many further references are found in each of the cited publications.

Chapter 1

W. Kleber, *An Introduction to Crystallography*, VEB Verlag Technik, Berlin (1971).

C. S. Smith, *Macroscopic Symmetry and Properties of Crystals*, vol. 6 of *Solid State Physics*, F. Seitz and D. Turnbull, Eds., Academic Press, New York (1958).

W. A. Wooster, *A Text-Book on Crystal Physics*, Cambridge Univ. Press, Cambridge, (1949).

Chapter 3

F. G. Fumi, "Physical Properties of Crystals: The Direct Inspection Method," *Acta Cryst.* **5**, 44 (1952).

F. G. Fumi, "The Direct-Inspection Method in Systems with a Principal Axis of Symmetry," *Acta Cryst.* **5**, 691 (1952).

C. Hermann, "Tensoren und Kristallsymmetrie," *Z. Kristall.* **A89**, 32 (1934).

H. J. Juretschke, "Transformation Properties of the Physical Constants of Crystals," *Acta Cryst.* **5**, 148 (1952).

H. Wondratschek, Ueber Tensorsymmetrien in den einzelnen Kristallklassen, *N. Jahrbuch Mineral.* **8**, 217 (1952).

Chapter 5

V. Heine, "The Thermodynamics of Bodies in Static Electromagnetic Fields," *Proc. Cambridge Phil. Soc.* **52**, 546 (1956).

H. A. Leupold, "Notes on the Thermodynamics of Dielectrics," *Am. J. Phys.* **39**, 1099 (1971).

J. A. Osborn, "Demagnetizing Factors of the General Ellipsoid," *Phys. Rev.* **67**, 351 (1945).

Chapter 6

W. F. Brown, Jr., R. M. Hornreich, and S. Shtrikman, "Upper Bound on the Magnetoelectric Susceptibility," *Phys. Rev.* **168**, 574 (1968).

T. H. O'Dell, *The Electrodynamics of Magnetoelectric Media*, American Elsevier, New York (1970).

B. A. Tavger and V. M. Zaitsev, "Magnetic Symmetry of Crystals," *Sov. Phys. JETP* **3**, 430 (1956).

Chapter 7

H. J. Juretschke, "Symmetry of Galvanomagnetic Effects in Antimony," *Acta Cryst.* **8**, 716 (1955).

W. H. Kleiner, "Space-Time Symmetry of Transport Coefficients," *Phys. Rev.* **142**, 318 (1966).

P. V. Pantulu and E. Sudarshan, "Magnetic Symmetry and Transport Properties of Crystals," *Acta Cryst.* **A26**, 163 (1970).

V. G. Shavrov and E. A. Turov, "Galvanomagnetic Effects in Ferrimagnetics near the Compensation Point," *J. Exptl. Theoret. Phys. (USSR)* **45**, 349 (1963).

S. Shtrikman and H. Thomas, "Remarks on Linear Magneto-Resistance and Magneto-Heat Conductivity," *Solid State Commun.* **3**, 147 (1965).

L. J. Van der Pauw, "Determination of Resistivity Tensor and Hall Tensor of Anisotropic Conductors," *Philips Res. Reports* **16**, 187 (1961).

J. D. Wasscher, "Note on Four-Point Resistivity Measurements on Anisotropic Conductors," *Philips Res. Reports* **16**, 301 (1961).

R. F. Wick, "Solution of the Field Problem of the Germanium Gyrator," *J. Appl. Phys.* **25**, 741 (1954).

Chapter 8

C. A. Domenicali, "Irreversible Thermodynamics of Thermoelectricity," *Rev. Mod. Phys.* **26**, 237 (1954).

W. H. Lucke, *A Brief Survey of Elementary Thermoelectric Theory*, U. S. Naval Research Laboratory Report 5888 (1963).

R. Wolfe and G. E. Smith, "Experimental Verification of the Kelvin Relation of Thermoelectricity in a Magnetic Field," *Phys. Rev.* **129**, 1086 (1963)

Chapter 10

G. C. Baldwin, *An Introduction to Nonlinear Optics*, Plenum Press, New York, (1969).

D. S. Chemla and J. Jerphagon, "Optical Second Harmonic Generation in Paratellurite and Kleinman's Symmetry Relations," *Appl. Phys. Letters* **20**, 222 (1972).

D. A. Kleinman, "Nonlinear Dielectric Polarization in Optical Media," *Phys. Rev.* **126**, 1977 (1962).

S. H. Wemple and M. DiDomenico, Jr., *Electrooptical and Nonlinear Optical Properties of Crystals*, vol. 3 of *Applied Solid State Science*, R. Wolfe, Ed., Academic Press, New York (1972).

Chapter 11

F. E. Borgnis, "Specific Directions of Longitudinal Wave Propagation in Anisotropic Media," *Phys. Rev.* **98**, 1000 (1955).

H. B. Huntington, *The Elastic Constants of Crystals,* vol. 7 of *Solid State Physics,* F. Seitz and D. Turnbull, Eds., Academic Press, New York (1958).

M. J. P. Musgrave, "Propagation of Elastic Waves in Aeolotropic Media I, II," *Proc. Roy. Soc.* **A226**, 339 (1954).

A. S. Pine, "Direct Observation of Acoustical Activity in α-Quartz," *Phys. Rev.* **B2**, 2049 (1970).

Chapter 12

A. R. Hutson and Donald L. White, "Elastic Wave Propagation in Piezoelectric Semiconductors," *J. Appl. Phys.* **33**, 40 (1962).

J. F. Scott, "Light Scattering from Polaritons," *Am. J. Phys.* **39**, 1360 (1971).

Chapter 13

H. J. Bunge, *Mathematische Methoden der Texturanalyse,* Akademie-Verlag, Berlin (1969).

L. H. Grabner and J. A. Swanson, "Symmetry Restrictions on Field-Dependent Tensors with Application to Galvanomagnetic Effects," *J. Math. Phys.* **3**, 1050 (1962).

Z. Hashin and S. Shtrikman, "Conductivity of Polycrystals," *Phys. Rev.* **130**, 129 (1963).

R. A. McDonald, "Cauchy Relations for Second and Third Order Elastic Constants," *Phys. Rev.* **B5**, 4139 (1972).

Index

Anisotropy, 1

Bismuth, 100

Character, 38
  antisymmetric product representations, 41, 204
  crystal group, 42
  irreducible decomposition, 44, 174, 177, 202
  irreducible representations of point groups, 42, 205–209
  irreducible representations of rotation group, 38, 39, 200
  orthogonality, 43
  principle of composition, 39
  symmetric product representations, 41, 203
  tensor transformations, 39–41, 201–204
Coordinate system, 18–19
  basis vectors, 18
  orthogonal axes, 18
  of scaled isotropic medium, 58, 76
  sense, 19
  of tensor components group, 34–37
Crystal, 4
  basis, 4
  crystal structure, 4
  lattice, 4
  translational symmetry, 4
Crystal groups, 9–10, 191–193
  group properties of symmetry operations, 6
  Hermann–Mauguin notation, 191
  Schönflies notation, 191
Crystal lattices, 4–15
  body-centered cubic, 13
  face-centered cubic, 13

Crystal lattices, (*cont'd.*)
  simple cubic, 13
  in three dimensions, 10–14
  in two dimensions, 5–8
Crystal optics, 104–133
  biaxial crystals, 111–113
  conical refraction, 113
  free-energy formulation, 128
  dispersion of optic axes, 113
  double refraction, 114
  optic axis, 111–112
  second order effects, 120–131
  uniaxial crystals, 108–111
Crystal systems, 10–14
  cubic, 13
  hexagonal, 12
  monoclinic, 11
  orthorhombic, 11
  rhombohedral, 12
  tetragonal, 12
  triclinic, 10
Cylindrical symmetry
  axial invariants, 45
  axially polycrystalline materials, 182–185
  construction of tensors of rank higher than six, 175
  number of independent tensor components, 174

Dielectric constant tensor, 49
  modification in piezoelectric media, 161
  symmetry in crystal groups, 50–51
Diffusion, 71
Direct inspection method for constructing tensors in crystals, 25

Elastic constants, 137–140

Elastic constants, (cont'd.)
  adiabatic and isothermal, 144–145
  Cauchy relations, 185
  dependence on electric field, 172
  dependence on magnetization, 173
  determination by static methods, 142–144
  elastic moduli (stiffness constants), 137
  elastic constants (compliance constants) 137
  fourth order, 173, 189
  of isotropic material, 141–142
  of polycrystalline material, 140–141, 177–182, 189
  third order, 172, 177
  two index notation, 136, 138
  transformation properties, 137–140
Elasticity, 134–153
  elastic strain, 135
  elastic stress, 135
  Hooke's law, 136
  homogeneous deformation, 134
  in piezoelectric media, 155
Elastic waves, 145–147
  acoustic and optical branches of spectrum, 162
  in piezoelectric crystals, 158–160
  mode amplitudes, 147
  polaritons, 161
  secular determinant for velocities, 146
  triple refraction, 147
Electric polarization, 48–60
  depolarization field, 52–54
  electric susceptibility, 17, 48
  force on dielectric ellipsoids, 57
  free energy in electrostatics, 50
  measurement of dielectric constants, 54–57
  potential distributions, 57–58
  shape anisotropy, 53–54
  shape depolarization tensor, 53
  torque on dielectric ellipsoids, 56
Electrical conduction, 71–90
  anisotropic conduction, 74–76
  current distribution of point source, 77
  potential distribution in anisotropic conductors, 76–79
  transport in magnetic field, 81–86
  transport in magnetic materials, 86–87
Electrical conductivity, 71, 74
  four-probe measurement, 79–81, 89
  isothermal, 93
  scale transformation in measurement of, 76
  symmetry of tensor, 74
Electrical resistivity (see also Electrical conductivity), 74
  measurement of, 74
  symmetry of tensor, 74

Electromagnetic waves in crystals, 104–118
  boundary conditions of fields, 114
  direction of energy propagation, 107
  excitation of modes, 113
  Fresnel formulas, 106
  group velocity, 110, 115
  in piezoelectric media, 160–161
  intrinsic symmetry of dielectric constant, 106
  phase velocity surfaces, 104, 109–110, 112–113
  Poynting vector, 106
  principal axis velocities, 107
Electrochemical potential, 92
Electrooptic effect, 127–128, 132

Faraday effect (see also Crystal optics), 127
  symmetry as second rank tensor, 127

Group representations
  irreducible, 38
  reducible, 39
  of point groups, 205–209
  symmetric product, 40

Hall effect, 17, 81–86
  Corbino disk, 85
  extraordinary, 87
  Hall conductivity, 84
  isotropic medium, 84
  potential distributions with, 84
  symmetry, 82
  ordinary, 81–86, 87
  third order, 171
Hooke's law, 136–137, 169

Invariants
  cylindrically symmetric tensors, 182
  determination of number of tensor invariants, 43–44, 174
  elastic tensors, 32–34, 141–142, 176
  linearly independent invariants, 30
  linear rotational invariants of tensors, 29–31, 175
  scalars and, 30
  subspaces of tensor components, 34
  of tensors, 29–33
  tensor decomposition into invariant subspaces, 34–38
  of transformations, 38
Irreversible processes (see also Thermoelectricity), 71–103
  entropy production in, 72, 92
  generalized conductivity of, 72
  Joule heat, 94
  microscopic reversibility and, 73

Index

Irreversible processes (*cont'd.*)
  Onsager relations, 17, 72, 73, 81, 86, 91, 93
  symmetry, 71-74
Isotropic materials (*see also* Polycrystalline media)
  elastic properties, 140-142
  form of second rank tensor, 30
  fourth rank tensors, 33
  higher order tensors, 174
  third rank tensor in rotational isotropy, 30

Magnetic conductors, 86-87
  linear magnetoresistance, 87
  ordinary and extraordinary galvanomagnetic effects, 87
  thermoelectricity, 102
  time reversal symmetry, 87
Magnetic anisotropy energy, 69
  fourth-order, 170
  sixth-order, 173
Magnetic polarization, 58
Magnetic susceptibility, 58
Magnetic symmetry, 61-70
  magnetic point groups, 65-66, 210-213
  time reversal and magnetic vectors, 61-63
Magneto-conductivity, 82, 171
Magnetostriction, 173
Magnetoelectric effect, 68
Matrix multiplication, 21

Onsager relations, 17, 72, 73, 81, 86, 91, 93
Optics *see* Crystal Optics
Optical activity, 120-126, 129
  circular polarization along optic axes, 126
  dispersion relations, 123
  intrinsic symmetry, 121
  normal modes, 125, 126
  as second rank tensor, 121
Optical phonons, 161

Photoelastic tensor, 170, 186
Piezoconductance, 186
Piezoelectricity, 154-168
  thermodynamics, 154-157
  parameters, 154-157
  redefinition of elastic and electric parameters, 156
  second-order, 188
Piezoelectric tensor, 155
  symmetry, 157
  transformation properties, 157-158
Piezomagnetism, 158
Polaritons, 161-165

Polaritons, (*cont'd.*)
  dispersion relations, 164
  mode parameters, 163
Polycrystalline media, 1, 140-142, 175-176
  relation of constants to single crystal constants, 142, 149, 181, 184, 189-190
Pyroelectric effect, 155, 188
Pyromagnetism, 66-67

Rotational symmetry, 1
  invariant subspaces of rotation group, 38, 196

Scale transformation in anisotropic potential distributions, 58, 76-79
Shape depolarization tensor of ellipsoids, 53
  effective susceptibilities, 54
  shape principal axis system, 53, 54
Symmetry
  anisotropy, 1
  isotropy, 1
  reciprocity relations, 2
  conservation laws, 2
  inherent symmetry, 2, 17
  point group symmetry, 5
  time inversion, 17
  of physical properties, 16, 17
  magnetic crystals, 61-70
Symmetry element, 4
  complete specification, 191-192
  inversion, 10
  order of element, 6
  rotation, reflection, 4, 6
  time reversal, 62-63
Symmetry group, 8, 191
  classes of elements, 43, 192
  generating elements, 8, 24, 194, 195
  multiplication table, 8
  point groups in two dimensions, 9-10
  subgroups, 8
  symmetry operations, 192
  32 point groups, 13

Tensor, 22
  axial tensor, 22
  determination of independent constants, 26
  orthogonality of linear combinations of tensor components, 32
  polar, 22
  of rank m, 22
  second rank, 22
  identification of index pairs with single indices, 83
  symmetry in isotropic materials (*see also* Invariants), 31
  transformation of, 21

Thermal conductivity, 71, 93–94
Thermal expansion tensor, 145
Thermoelectric effects, 91
  relations between, 99–100
Thermoelectricity, 91–103
  Bridgman effect, 95, 97, 98
  Peltier effect, 95–97
  Seebeck effect, 98–99
  sources of heat generation, 94
  thermocouple, 99
  Thomson effect, 95, 98
Thermoelectric tensors, 93
  in magnetic fields, 186
  symmetry, 93, 94
Transformation, 18–24
  of basis vectors, 19
  block form of transformation matrices, 36

Transformation, (*cont'd.*)
  consecutive, 20
  covariant and contravariant, 22
  fixed point with respect to new axes, 20
  generalized coordinate transformation, 34
  orthogonal, of coordinates, 18
  practical rule for carrying out tensor transformations, 22
  tensors, 21

Vector, 22
  axial, 22
  magnetic, 65
  polar, 22
  representing tensor components, 34–38